# 行走上海美术馆
## ROAMING SHANGHAI'S ART MUSEUMS
### A CityWalk Exploration

潘丽 著
Pan Li

同济大学出版社·上海
TONGJI UNIVERSITY PRESS · SHANGHAI

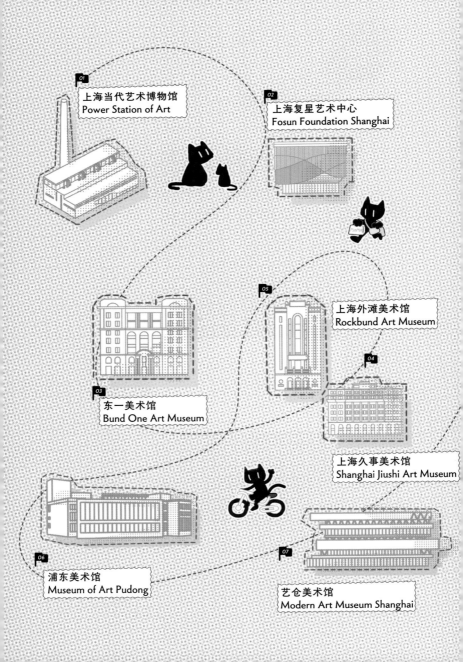

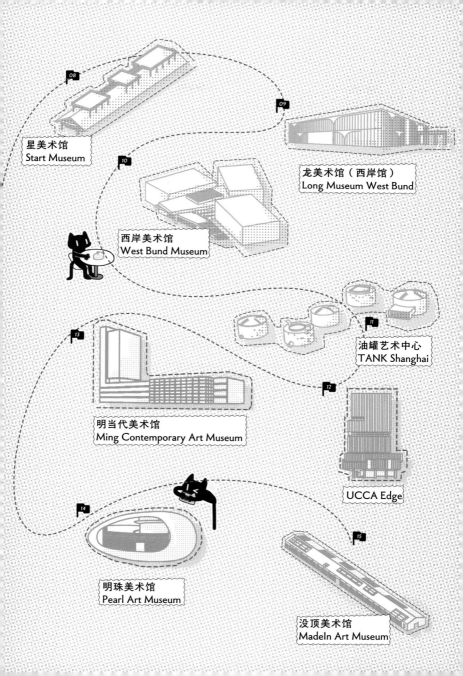

图片提供：同济原作设计工作室 Courtesy TJAD Original Design Studio

# 目录

前言 ……8

## 黄浦滨江&外滩沿线 ……14

- 01 上海当代艺术博物馆 ……16
- 02 上海复星艺术中心 ……36
- 03 东一美术馆 ……44
- 04 上海久事美术馆 ……60
- 05 上海外滩美术馆 ……68

## 西岸美术馆大道 ……110

- 08 星美术馆 ……112
- 09 龙美术馆（西岸馆） ……130
- 10 西岸美术馆 ……146
- 11 油罐艺术中心 ……160

## 滨江大道沿线 ……86

- 06 浦东美术馆 ……88
- 07 艺仓美术馆 ……102

## 其他区域 ……168

- 12 UCCA Edge ……170
- 13 明当代美术馆 ……178
- 14 明珠美术馆 ……186
- 15 没顶美术馆 ……194

编后记 ……202

# CONTENTS

PREFACE ......8

| HUANGPU WATERFRONT & THE BUND | ......14 |

- 01 Power Station of Art ......16
- 02 Fosun Foundation Shanghai ......36
- 03 Bund One Art Museum ......44
- 04 Shanghai Jiushi Art Museum ......60
- 05 Rockbund Art Museum ......68

| WEST BUND MUSEUM MILE | ......110 |

- 08 Start Museum ......112
- 09 Long Museum West Bund ......130
- 10 West Bund Museum ......146
- 11 TANK Shanghai ......160

| BINJIANG AVENUE | ......86 |

- 06 Museum of Art Pudong ......88
- 07 Modern Art Museum Shanghai ......102

| OTHER AREAS | ......168 |

- 12 UCCA Edge ......170
- 13 Ming Contemporary Art Museum ......178
- 14 Pearl Art Museum ......186
- 15 Madeln Art Museum ......194

AFTERWORD ......202

\*本书中各美术馆的开放时间，或有变动，请以各馆官网即时发布为准。

*The opening hours of the museums may change, please refer to the latest information via official websites/social media before your visit.*

抵达美术馆
即抵达一种想象
把自身从现实维度中释放
从日常中释放
抵达更多样的文明和生命体验

A Museum Visit Is:
An Extraordinary Opportunity for *Imagination*
*Liberation* from Mundane Routines of Daily Life
Opening the Door to a World of *Diversified Perspectives*

图片提供：UCCA尤伦斯当代艺术中心 *Courtesy UCCA Center for Contemporary Art*

# Preface

In the last two decades, an artistic network has flourished along the scenic banks of Shanghai's Huangpu River and Suzhou Creek, both prominent waterways in the city. As of 2023, the 6.3-kilometer waterfront along Suzhou Creek has been transformed into an awe-inspiring canvas housing more than 100 vibrant art spaces. Meanwhile, the Huangpu River has become a hub of artistic expression, featuring renowned cultural areas like the Bund, the "West Bund Cultural Corridor" project, initiated in 2010, and the post-Expo venues.

Shanghai's artistic tale revolves around water, guiding our exploration of the city in this book. Charting the art museums along the River and the Creek, we reimagine Shanghai's art landscape through the concept of a "water system."

The artistic journey along the Suzhou Creek art belt begins with the iconic M50 Creative Park. Back in 1999, the former Chunming Roving Factory at No. 50 Moganshan Road ceased production and underwent a remarkable transformation into the Chunming Urban Industrial Park. The availability of affordable rented factory spaces quickly lured artists, who soon set up their studios, and galleries like BizArt (co-founded by artist Xu Zhen and others) and ShanghART Gallery (founded by Lorenz Helbling) followed suit. The early days of M50 laid the very foundation for Shanghai's flourishing contemporary art ecosystem, and the artists and gallery owners who once thrived here have since become influential pillars of the Shanghai art scene.

During China's golden decade of rapid contemporary art development from the 2000s to the 2010s, the M50 art district blossomed organically. As the years progressed, the Suzhou Creek area witnessed further growth, with the establishment of prominent art museums like OCAT Shanghai (temporarily closed in 2023) and UCCA Edge. These additions formed a captivating artistic network along the Creek, solidifying its position as a vibrant hub for creativity and cultural expression in Shanghai.

As you take a leisurely stroll to the east end of Suzhou

# 前言

一江一河，黄浦江、苏州河作为上海的主要水脉，近20余年来发展孕育出一条傍水而生的艺术脉络。苏州河全长约6.3公里的滨水空间，至2023年已分布了100多处大小艺术空间。而黄浦江两岸更是汇聚了多重艺术地带：历史文化街区的外滩、2010年后系统开发的"西岸文化走廊"工程项目、"后世博"场馆等。

可以说，上海的艺术叙事是一场水的叙事，而本书的行走路径也是水。我们梳理了一江一河沿线的美术馆，用"水系"这一地理学的概念重新排演了上海的艺术版图。

苏州河艺术带的开端是M50。1999年位于上海莫干山路50号的春明粗纺厂（前身为创办于1937年的信和纱厂）停产，转型成为"春明都市产业园"。当时大量租金便宜的厂房吸引了艺术家入驻，被用作其工作室；然后是画廊入驻，比如比翼艺术中心（艺术家徐震等人创办）、香格纳画廊（劳伦斯·何浦林创办）。早期的M50奠定了上海当代艺术生态的肌理，彼时在此活跃的艺术家、画廊主，如今仍是上海艺术界最具活力、影响力的中流砥柱。

M50是自发形成的艺术区，这背后是中国当代艺术迅猛勃兴的黄金十年。2010年后，美术馆级别的艺术机构：OCAT上海馆（2023年暂时关闭）、UCCA Edge，相继入驻苏州河沿线，一条苏州河艺术脉络由此勾连，从苏州河东段至外白渡桥，抵达外滩。外滩的百年建筑变身美术馆，又生成了独具风貌的外滩艺术生态。

西岸是规划生成的艺术区。2011年年末徐汇区政府在其第九次党代会上明确提出打造"西岸文化走廊"的品牌工程战略之后，"上海西岸"这个新的文化地标正式浮出水面。[1] 西岸片区涵盖西岸艺术中心、西岸文化艺术示范区、

---

1 西岸官网，《巴黎有左岸，上海有西岸》（2016），
http://www.westbund.com/cn/index/NEWS-CENTER/westbund-leftbank.html

Creek, the iconic Garden Bridge awaits, leading you to the Bund, where a mesmerizing transformation has unfolded. Witness the magical rejuvenation of several century-old buildings, thoughtfully repurposed into captivating art museums, giving birth to a one-of-a-kind and vibrant Bund art ecosystem. The harmonious blend of heritage and contemporary art paints an enchanting ambiance, welcoming visitors to immerse themselves in the rich cultural tapestry of Shanghai's past and present. Within the walls of these splendid art institutions, an extraordinary experience awaits, capturing the essence of this dynamic city's artistic spirit.

Nowadays, the West Bund area in Shanghai has evolved into a thriving art district, known as "Shanghai West Bund." It emerged as a significant cultural landmark after the proposal to establish the "West Bund Cultural Corridor" during the ninth Party Congress of Xuhui District in late 2011. [1] The area encompasses various transformative projects, such as the West Bund Art Center, the West Bund Culture and Art Pilot Zone, the Shanghai West Bund Fine Art Storage, and the West Bund Museum Mile. Along this impressive Museum Mile, visitors encounter an awe-inspiring cluster of museums, including the Start Museum, the Long Museum (West Bund), the West Bund Art Center, and the TANK Shanghai, seamlessly integrated with waterfront green spaces and parks. These vibrant public areas have become beloved destinations, offering an enchanting artistic experience amidst Shanghai's picturesque setting.

The exploration of Shanghai's museums is intricately woven with the fascinating "history of museums." Liu Yichun, the architect of the Long Museum (West Bund), shared insights with us, revealing that museums began as private collections of aristocrats and nobles, inaccessible to the public. It was only later that the concept of museums opened up to the public domain. The pioneering modern museum in China, the Shanghai Museum, once resided in the former Royal Asiatic Society (RAS) building, now home to the Rockbund Art Museum. This transformative step marked the dawn of modern narration and heralded the advent of museums accessible to all, inviting people from all walks of life to marvel at the wonders of art and history.

Shanghai proudly holds the title of being the city with the highest number of art museums in China, featuring an impressive collection of 15 renowned institutions showcased

---

[1] "Paris has the Left Bank, Shanghai has the West Bund," Interview with Chen Anda, *Cultural Fortune*, 2016. http://www.westbund.com/cn/index/NEWS-CENTER/westbund-leftbank.html (in Chinese)

西岸艺术品保税仓库、西岸美术馆大道等诸多项目。美术馆大道自北向南分布着——星美术馆、龙美术馆、西岸美术馆、上海油罐艺术中心,形成颇为壮观的博物馆集群,并与滨江绿地和公园一起共生,成为颇受大众喜爱的公共区域。

依然是与"水"相关。以"城"为主题的2010年上海世博会选址在黄浦江两侧。世博会结束后,"中国馆"成为中华艺术宫,延续老上海美术馆的职能;"城市未来馆"则改造为上海当代艺术博物馆。

当我们以水为脉串联起上海的美术馆时,"博物馆史"这一时间脉络也蕴含在其中。建筑师柳亦春在设计龙美术馆(西岸馆)时做过博物馆考据:作为公共概念的博物馆在步入现代生活之前,是以私人的面貌存在的。博物馆的雏形是皇家、贵族的私人珍藏,它并不属于公众,也不对公众开放。位于外滩源的上海外滩美术馆,其建筑的前身亚洲文会大楼曾为上海博物院——中国最早成立的博物馆之一,也是近代中国第一个向公众开放的博物馆。对公众开放,是现代叙事的开始,也是博物馆的开始。

上海是全国拥有最多美术馆的城市。本书中介绍的15座美术馆涉及不同的运营模式:有公立的,有私人藏家的,也有民营公司的。虽运营背景各不相同,但现代美术馆在建构时都会考虑到"公共性"——无论是建筑景观与公共休闲区的连通、共建,还是向公共免费开放的展览、美术馆的免费开放日,或是精心策划的公共教育活动。将美术馆的"围墙"拆除,让其融入城市生活,不仅是建筑师在建造美术馆之初的理念,也是各座美术馆在各项展览和活动中的实际行动。

这是一本写给公众的美术馆手册。每一条抵达美术馆的路径,我都反复走过。比如,从星美术馆到龙美术馆(西岸馆),有三种步行路径:沿着滨江走、沿着瑞宁路走、沿着铁轨走。抵达美术馆,即抵达一种想象,把自身从现实维度中释放、从日常中释放,抵达更多样的文明和生命体验。

in this book. While this selection may not encompass the entirety of Shanghai's museum landscape, it provides a broad representation of various operational models, including public, privately-owned, and privately financed institutions. Despite their differences, all modern art museums share a common commitment to public engagement. This is evident through architectural integration with public leisure spaces, free exhibitions, open visiting days, and diverse educational programs. The overarching goal is to break down the traditional barriers of museums, seamlessly blending them into urban life. This approach is not merely an architectural concept; it serves as a guiding principle in the curation of exhibitions and the design of engaging activities, nurturing a dynamic and accessible artistic experience for all.

This book serves as the ultimate guide for museum visitors, leading you through every path that leads to these captivating institutions. A museum visit is an extraordinary opportunity for imagination, as it liberates you from the mundane routines of daily life, opening the door to a world of diversified perspectives. Within these hallowed halls, you become an integral part of the museum experience; your presence completes the process of observing, interacting, and participating in the artworks, transforming you into a creator of culture. Embrace the journey of artistic exploration, where each museum visit becomes a transformative and enriching encounter with creativity and human expression.

*Pan Li*
*August 2023*

而你,也会成为美术馆的一部分;通过你的抵达,完成作品的观看、行动、参与,成为文化的创作者。

潘丽

2023 年 8 月

# 黄浦滨江 & 外滩沿线
# HUANGPU WATERFRONT & THE BUND

The riverside area of a city has always been a nurturing ground for the art industry, offering urban residents an ideal place for cultural and leisure activities. Our city walk begins at the Huangpu waterfront, also known as Huangpu Expo Binjiang, where modern industry in Shanghai was born. The transformation of the former power plant into an art museum has infused this area with a strong artistic essence and seamlessly linked it to the artistic atmosphere along the Bund.

Heading north and crossing the Nanpu Bridge, we reach the Bund area—a renowned historical and cultural district along the Huangpu River. The Bund marks the starting point of Shanghai's modern urban development and plays a significant role as the birthplace of modern Chinese art. Various forms of contemporary art, such as art publications, cultural collections, film production, commercial advertising, and photography exhibitions, all originated from the Bund and continue to drive the city's artistic vibrancy. Since 2010, the Bund area has gradually attracted numerous art institutions, including art museums, galleries, and auction houses, giving rise to a naturally evolving "Bund art belt." This harmoniously blends with the old alleys, the historical architecture cluster, and the bustling crowds of tourists, creating a captivating new landscape on the banks of the Huangpu River.

滨江空间历来是艺术产业蓬勃发展的风水宝地，也为城市居民提供了文化和休闲活动的理想场所。我们的"行走上海美术馆"，首先从黄浦滨江开始。这里又被称为黄浦世博滨江，也是上海近代工业的发源地。"电厂"的"重新发电"为这片区域注入了强劲的艺术基因，也使其更好地与外滩沿线的艺术氛围做了衔接。

从黄浦滨江一路向北，穿过南浦大桥就到了外滩片区。外滩是黄浦江畔的著名历史文化街区，它是上海近代城市发展的起点，更是中国现代艺术的重要发源地。艺术出版、文博典藏、电影制作、商业广告与摄影展……这些现代艺术形式正是以外滩为起点，进而引领着整座城市艺术生活向前发展。2010年后，外滩地区逐渐集聚了数十家艺术机构，含美术馆、画廊、拍卖行等，构成了一片自然形成的"外滩艺术带"，与老弄堂、万国建筑博览群、如织的游人融合，成为一道新的外滩风景。

上海当代艺术博物馆全景　Panorama, Power Station of Art
图片提供：上海当代艺术博物馆 *Courtesy Power Station of Art*

# 01
# 上海当代艺术博物馆
# Power Station of Art

建筑师：同济原作设计工作室 ｜ 开馆时间：2012 年
Architects: TJAD Original Design Studio ｜ Opening Year: 2012

> PSA不仅对上海这座城市而言意义非凡，
> 它也是中国当代艺术的"发电"现场。
>
> The Power Station of Art plays a pivotal role as both
> a "power generator" for the city of Shanghai and a
> driving force for contemporary art in China.

　　上海当代艺术博物馆（简称PSA）坐落于南浦大桥边的黄浦滨江沿岸，其建筑前身是南市发电厂，人们亲切地称之为"电厂"。原电厂的附属设施、高达165米的大烟囱也转身为PSA的标志。

　　由电厂转身为美术馆，建筑空间经历了9个月的改造。设计团队秉持"有限干预"的设计精神：尊重原有空间的秩序及精神，最大限度让厂房的外部形态与内部空间的原有秩序和工业遗迹特征得以体现——保留了电厂标志性的烟囱、发电机、高中低三级梯度的厂房空间，以及屋顶上四个巨大的煤粉分离器，贯穿整座建筑的流线组织系统也借鉴了输煤栈桥的形态意象。输煤栈桥弯折如Z字形的游走路径，演绎成为建筑空间中的楼梯、扶梯和廊道，打破了层与层之间、室内室外之间的界限，让楼梯"弥漫"在随时需要的地方；输煤栈桥的形态意象也强烈如标识一般镶嵌进建筑的形体和空间。

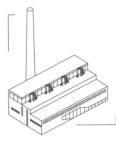

The Power Station of Art (PSA), situated along the Huangpu waterfront near Nanpu Bridge, earns the endearing nickname of the "power plant" due to its previous identity. The most notable feature of this former power plant is its iconic 165-meter-high chimney, which served as a prominent symbol of the Nanshi Power Plant.

Led by TJAD Original Design Studio, the renovation spanned nine months with a clear objective of preserving the building's original spatial arrangement and industrial essence through carefully managed interventions. The project successfully retained the iconic chimney, generators, and the distinct three-level layout of the factory spaces. Additionally, the four imposing fly ash separators located on the roof were meticulously preserved, maintaining their historical significance within the overall design of the renovated space.

Taking inspiration from the industrial heritage of the Power Station of Art, the circulation system within the building was ingeniously designed to emulate the characteristics of a coal conveyor bridge. Through this architectural reinterpretation, the bridge elements were transformed into functional components such as stairs, escalators, and corridors, seamlessly integrated into the spatial layout. This innovative approach breaks the traditional boundaries between floors and blurs the distinction between indoor and outdoor spaces, allowing stairs to appear wherever they are needed, creating a fluid and dynamic environment. The essence of the coal conveyor bridge is artfully embedded within the building's form and spatial organization, serving as a prominent feature. Once responsible for conveying coal to generate electricity, it now fulfills a new purpose as a conduit for artistic nourishment, welcoming numerous visitors into a vibrant and engaging art scene.

The project architect expressed her reflection on this transformation in the interview, stating, "The renovation process witnessed the metamorphosis of the building from a colossal electrical apparatus into a dynamic engine that promotes culture and art. It is akin to a historical narrative, encapsulating the condensed history of urban transformation within the Power Station of Art. In this profound undertaking, the significance of architectural practice extends beyond an individual building or architectural typology. It reflects the enduring imprint of a city's determined growth and stands as a testament to the persistent contemplation of its practitioners."

The reinterpretation of the original space within the Power Station of Art goes beyond spatial design and serves as a vital element in the artistic narrative of the museum. Shortly after its opening in 2012, PSA hosted the 9th Shanghai Biennale, which revolved around the theme of "Reactivation." This theme resonated

输煤栈桥弯折如Z字形的游走路径,演绎成为建筑空间中的楼梯、扶梯和廊道(摄于2012年)
The coal conveyor bridge was reimagined and transformed into components such as stairs, escalators and corridors (shot in 2012)
图片提供:同济原作设计工作室 *Courtesy TJAD Original Design Studio*

deeply with the venue's former identity as a power plant, symbolizing its historical significance as the backbone of China's industrial heritage. Once responsible for the transformation of resources, the power plant has undergone a remarkable metamorphosis, evolving into an energetic engine that drives contemporary art. This transformation not only reflects the adaptive reuse of a post-industrial space but also highlights the power of art to repurpose and breathe new life into historical contexts. PSA stands as a testament to the creative potential and cultural reactivation that can arise from the revitalization of industrial spaces.

The colossal chimney of PSA, once an integral part of the power plant, has been transformed into a site where captivating art installations unfold. Over the years, it has witnessed remarkable artistic expressions. In 2012, Roman Signer's "Ball with Blue Paint" dropped from the chimney's height, leaving a vibrant mark on the floor. Christian Boltanski's "Hearts" in 2018 synchronized heartbeats with a hanging light bulb, casting a luminous glow in harmony with the amplified heartbeats. Pablo Vargas Lugo's "Eclipses for Shanghai" represented future eclipses through animated cards and orchestral music, involving 125 high school students. In 2021, Carlos Casas utilized the chimney as a "geomagnetic electro-acoustic instrument" in "1883/Krakatoa," resonating with the sound of the Krakatoa eruption. These installations showcase the versatility and artistic potential of the chimney, creating immersive experiences and adding to the dynamic art scene at PSA.

The first floor of PSA is a grand space with an expansive area of approximately 960 square meters. The awe-inspiring atrium spans across five floors and reaches a towering height of 20.8 meters. This extraordinary atrium serves as a captivating exhibition hall, regularly showcasing visually stunning artworks that become the centerpiece of themed exhibitions. In 2014, visitors to the Great Hall were greeted by a captivating installation titled "The Ninth Wave" by artist Cai Guo-Qiang. This artwork featured 99 fabricated animals aboard a fishing boat that embarked on a symbolic journey from the artist's hometown of Quanzhou. Navigating along the Huangpu River, passing by the iconic Bund, and finally arriving at the Power Station of Art, the boat and its animal passengers evoked a powerful sense of movement and exploration. The animals, which included tigers, pandas, and camels, appeared weather-beaten and with bowed heads, symbolizing their endurance through the challenging and tumultuous currents of our times. This installation not only highlighted the artist's connection to his hometown but also served as a thought-provoking reflection on the global issues of migration, survival, and the interdependence of humans and animals.

In 2016, the exhibition hall featured a captivating artwork titled "Head" by Huang Yong Ping, a Chinese-born French avant-garde artist. This thought-provoking installation presented an 11-meter-long, 20-ton section of a green-skinned train suspended at a 30-degree angle. From the train cabin,

电厂时期输煤栈桥负责输送燃料；如今，它输送艺术养料，将南来北往的人送达最火热的艺术现场。

"改造历程以一种历史叙事的方式见证了昔日能源输出的庞大机器如何转变为推动艺术发展的巨大引擎。上海当代艺术博物馆的演变过程几乎就是一段浓缩的城市变迁史。在这个值得回溯的进程中，建筑实践的意义早已超越了建筑单体或建筑类型，从它身上可以辨识出一座城市顽强生长的印记以及一批实践者坚持思考的佐证。"参与项目的建筑师介绍道。

像是一种精神绵延，对原址空间的再叙事不仅体现在空间设计上，也成为美术馆艺术叙事的重要线索。2012年PSA甫一开馆便迎来了第9届上海双年展，"重新发电"的双年展主题便是对原址空间"电厂"的呼应。"电厂"曾是中国工业摇篮的命脉，承载了资源变革的使命；如今此地转变为当代艺术的能量发动机。原为电厂设置的大烟囱，也成为艺术发生的现场。2012年罗曼·辛格在大烟囱内部扔下了"第一弹"：一个装满蓝漆的球在30米高处被抛下，坠地瞬间，油漆四溅（作品《装蓝漆的球》）。此后，各国知名艺术家纷纷进入这个巨大、幽深、巨鲸身体一般的工业体内部进行创作：模拟人类心跳，制造日食、火山这类天文地质奇观。克里斯蒂安·波尔坦斯基在烟囱内部悬挂一盏大型工业灯泡，灯泡的闪烁频率与艺术家本人的心率同步（作品《心》）；巴勃罗·巴尔加斯·卢戈邀请125名上海中学生走进烟囱，以彩色纸板与管弦乐的方式演绎日食现象（作品《上海日食》）；艺术家卡洛斯·卡萨斯利用烟囱的特殊声学构造，将烟囱改造成一件"地磁电声乐器"，观众可以聆听印度尼西亚喀拉喀托火山爆发的声音（作品《1883/喀拉喀托》）……由此，烟囱也被人们称为PSA最奇特的展厅。

除了烟囱，PSA还有几处特别的展览空间。一层入口的大厅，挑高近20.8米、面积约936平方米，中间的天井贯穿建筑的一至五层。这个展厅承担着"亮相"功能，通常一个主题展中最重要、最具视觉震撼的作品会在此呈现。2014年，蔡国强的个展"九级浪"，一艘真实的渔船从艺术家的故乡泉州"乘风破浪"行至黄浦江，最终停泊于展厅，船上载着老虎、熊猫、骆驼等99只仿真动物，仿若诺亚方舟。不仅有渔船，还有火车、成吨的旧衣服，这个入口处总能"冲"入各种意想不到之物。2016年，一辆长达11米、重达20吨的半截绿皮火车车厢以倾斜30°角、被悬挂的姿态出现，而40只无头动物仿若从车厢中倾巢而出，四处奔逃。这是黄永砅（1954—2019）个展"蛇杖Ⅲ：左开道岔"的作品，这位善于哲思、于中国及世界都极为重要

"蔡国强：九级浪"展览开幕烟花秀，2014
Fireworks on the opening of "Cai Guo-Qiang: The Ninth Wave", 2014
图片提供：上海当代艺术博物馆
*Courtesy Power Station of Art*

"蔡国强：九级浪"展览现场，2014
Cai Guo-Qiang: The Ninth Wave, 2014
图片提供：上海当代艺术博物馆
*Courtesy Power Station of Art*

巴勃罗·巴尔加斯·卢戈的作品《上海日食》在大烟囱内演绎，2018
Pablo Vargas Lugo's work "Eclipses for Shangh was performed at the chimney in 2018
图片提供：上海当代艺术博物馆
*Courtesy Power Station of Art*

的艺术家希望用这些无头动物来提醒一种危机感。2018年,第二次世界大战后的欧洲重要艺术家克里斯蒂安·波尔坦斯基将10吨衣服放在PSA一层,高达15米的起重机钩爪悬挂在衣物堆砌而成的山峰之上,重复抓起、抛下衣服的动作。波尔坦斯基擅长选取日常物件作为主要创作材料,这些旧衣隐喻了人的肉身。

一层展厅的自动扶梯、楼梯,既是通往二层展厅的路径,也是展览叙事的一部分。梁绍基是一位隐居在浙江天台山的"蚕痴",他养蚕并通过蚕完成作品,在2021年9月开幕的个展中,他将PSA一层展厅的扶梯变身为作品《蚕道》:洁白的绡[1]笼罩于整个扶梯,轻盈、梦幻,阳光照耀其上,熠熠生辉;人穿行其中,如蚕进入茧,人如蚕。

另一处特别展厅是五层的临江平台。这个露天平台视野空阔,与自然相连,既可成为人们观展途中的休憩之所,也是一处户外展厅。西野康造的装置作品《与呼吸的声音》作为永久馆藏伫立在蓝天下,一对由金属钛制作的巨大又轻盈的蜻蜓之翅,在风中飘舞。

PSA不仅对上海这座城市而言意义非凡,它也是中国当代艺术的"发电"现场。始创于1996年的上海双年展,是中国最早创办的国际当代艺术双年展,不仅促发了中国当代艺术的勃兴,也成为亚洲最富影响力的艺术事件之一。上海双年展在发展了20年后,已由艺术圈、学术圈的内部事件转为城市文化事件,成为上海的一张城市名片。2021年第13届上海双年展的宣传短片《爱他,爱她,爱TA……》在各大平台播放,这条短片由艺术家杨福东拍摄,邀请了7位各个领域的名人出镜(作家孙甘露、学者刘擎、作家许知远、艺术家丁乙、表演艺术家谭卓、歌唱家黄英、音乐唱作人王嘉尔),讲述自己与上海的故事。

也是从PSA建馆那年开始,上海双年展开辟了一条"更上海"的叙事线,每一届策展人或策展团队都会立足"上海"开启主题阐释。第10届的主题"社会工厂"是基于上海"高速孕育后工业社会"的城市表征,以及20世纪20年代兴起于上海的中国木刻运动;第13届的主题"水体",是基于上海的地理及文化特征,上海水系发达、靠江临海,可以说是一座水做的城市,城市文化也具有"水"的包容和开放。关于上海的城市叙事,在PSA持续、系统地展开。自2012年起上海双年展还专门开辟了"城市项目",让展览、展映、田野调查、工作坊等与整个城市联动;走出美术馆围墙,并不仅是在物理层面打破空间的限制,更是在精神层面上做到与普通人相连——第11届

---

[1] 绡,一种半生半熟的丝。

forty headless animals appeared to pour out in different directions, creating an atmosphere of chaos and a sense of fleeing. Huang Yong Ping explained that the headless animals symbolized a profound sense of crisis within contemporary society. The artwork sparked contemplation on the challenges, uncertainties, and disruptions faced by individuals and communities in the modern world, inviting viewers to reflect on the complexities of our times and the consequences of societal upheaval.

In 2018, the Power Station of Art showcased a compelling installation titled "Personnes" as part of Christian Boltanski's solo show "Storage Memory." This artwork dominated the expansive space and featured a towering mountainous pile of garments weighing nearly 10 tons. Suspended above the garments, a 15-meter-tall crane repetitively and endlessly grabbed and released the clothes in a mesmerizing motion. Boltanski, a renowned French artist and influential figure in the post-WW II European art scene, used recycled human clothing as a poignant symbol of flesh and human existence. The crane, on the other hand, represented the hand of God and the inescapable grip of fate. This thought-provoking installation invited viewers to reflect on the fragility and transience of life, the vastness of collective memory, and the interconnectedness of human experiences. Through the juxtaposition of clothing, machinery, and movement, "Personnes" delved into profound existential questions and evoked a contemplative response from visitors.

The escalator and staircase leading to the upper floor of the Power Station of Art not only serve a functional purpose but also play a significant role in the exhibition narrative. In Liang Shaoji's solo show, "A Silky Entanglement" (September 2021–February 2022), the first floor escalator underwent a captivating transformation. Liang enveloped it in pure white raw silk, creating a cocoon-like tunnel that exuded an ethereal beauty and weightless ambiance. As a dedicated "silkworm enthusiast," Liang has immersed himself in the artistic practice involving silkworms for decades. With this installation, he invites viewers to step into his art world and experience the sensation of becoming a silkworm. Walking through the silk tunnel blurs the boundaries between human and insect, evoking a transformative and thought-provoking encounter. This immersive experience challenges conventional notions of existence and prompts contemplation on our interconnectedness with the natural world. By repurposing the escalator, Liang's installation transcends its utilitarian function, becoming a conduit for profound artistic exploration and inviting viewers to reflect on themes of metamorphosis, interconnectedness, and the delicate beauty of the silkworm's process.

The fifth-floor terrace at the Power Station of Art offers visitors a dual-purpose experience as both a serene relaxation area and an outdoor exhibition space. Amidst the expansive open space beneath the boundless blue sky, visitors can

潘岱静作品《重奏》(2021),利用了PSA大厅特有的挑高中庭空间,第13届上海双年展/上海当代艺术博物馆委任创作
"Done Duet" by Pan Daijing, 2021. Commissioned by the 13th Shanghai Biennale/Power Station of Art
图片提供:上海当代艺术博物馆 *Courtesy Power Station of Art*

behold the permanent installation artwork titled "Harmony with the Breeze" by Kozo Nishino. This captivating installation features a remarkable pair of colossal dragonfly wings meticulously crafted from titanium. The delicate and graceful design of these wings allows them to sway and flutter gently with the wind, infusing the surroundings with an air of elegance and dynamic movement. As visitors appreciate the artwork on the terrace, they are enveloped in a harmonious connection with nature, where the interplay of the artwork, the breeze, and the vast sky evokes a sense of tranquility and awe-inspiring beauty.

The Power Station of Art plays a pivotal role as both a "power generator" for the city of Shanghai and a driving force for contemporary art in China. A testament to this influence is the Shanghai Biennale, which was initiated in 1996 and stands as China's earliest international contemporary art biennial. Throughout its history, the Shanghai Biennale has played a crucial role in nurturing the growth of contemporary art within China and has emerged as one of the most influential art events in Asia. The biennial serves as a platform for artists, curators, and art enthusiasts to explore and showcase innovative artistic expressions, pushing boundaries, and fostering cultural dialogue.

Since 2012, PSA has assumed a pivotal role as the primary organizer and permanent exhibition venue for the Shanghai Biennale. This prestigious international art event has embraced a distinctive "Shanghai-oriented" narrative, with each edition's curatorial team deriving inspiration from the city itself. The

西野康造的装置作品《与呼吸的声音》作为永久馆藏伫立在PSA五层的临江平台上
"Harmony with the Breeze" by Kozo Nishino, PSA's permanent installation artwork on the fifth-floor terrace
图片提供：上海当代艺术博物馆 *Courtesy Power Station of Art*

双年展的外延项目"51人"计划,便征集了这座城市51位不同职业的劳动者,用51场在上海不同地点展开的公共活动,实现了双重层面的"走出美术馆"。

PSA专注于当代艺术的叙事,对中国当代活跃的艺术家做系统性的研究,同时也做国际艺术大师的展陈,并从零开始做馆藏。除了当代艺术,建筑也是PSA一直关注的方向。PSA馆长龚彦曾在采访中表示:"与当代艺术相较,建筑存在的意义是它和人之间的关系,它讲的是人,建筑本身就是一个生存装置;建筑师更像是一位人类社会的观察者,正视并挑战我们面临的问题。"[2] 自2014年起,PSA呈现了十余位世界级的建筑设计大师和工作室的展览,包括藤本壮介、让·努维尔、约翰·海杜克、赫斯维克工作室等,并对地域性、时代性的建筑思潮、现象展开系统研究。如"直行与迂回:台湾现代建筑的路径"展览梳理了20世纪40年代至今台湾现代建筑的发展脉络;"觉醒的现代性:毕业于宾夕法尼亚大学的中国第一代建筑师"展览则通过历史文献的呈现方式,以具体案例系统性梳理20世纪初期中国现代建筑的崛起与发展。业内建筑师曾评价:PSA是国内建筑展做得最好的美术馆。

在展示建筑的过程中,PSA也创造了新的建筑。2013年,馆方邀请设计师在美术馆的五楼打造了一个全新空间:在这个毗邻室外滨江露台的狭长空间内,白色马赛克铺陈出一片清凉静谧;而两处迷你"泳池",一圆一方,是颇具幽默感的设计,观众可以在此落座歇脚。新空间被命名为"SPA空间",这样的命名有着玩笑的气质,而泳池与马赛克又进一步把玩笑落实,非常高级的幽默。2015年,馆方邀请尤纳·弗莱德曼在PSA三层通往烟囱的连廊中,重建了他的"简单技术博物馆"项目。2016年,非常建筑工作室受邀完成了PSA的全新空间——上海当代艺术博物馆设计中心(psD)的设计,建筑师张永和希望,"这个博物馆建筑临街一侧的界面所勾勒的城市街道的写意,能够吸引人们通过这些'店面'进入psD,然后走向博物馆纵深。"[3]

psD专注于设计的展陈,由此除建筑设计之外,我们也看到了更多门类的设计呈现:服饰、珠宝、平面、家具等。有趣的是,当设计师进入美术馆,往往带着艺术家的野心;而设计作品从货架进入美术馆的排演系统时,也拥有了新的符号和解读。在psD首个与时尚相关的展"侯赛因·卡拉扬:群岛"中,我们看到这位服装设计专业出身的设计师,把身体和秀台作为表达的现场,以装置、影像等媒介去讨论移民、身份等议题,思想抵达之处早已超越了时装。

---

[2] 《龚彦:PSA成立十年,持续搭建"思想的塔"》,艺术新闻,2022
[3] "建筑之名:非常建筑泛设计展"导言,张永和/非常建筑,
https://www.powerstationofart.com/whats-on/exhibitions/the-name-of-architecture

thematic interpretations of the Shanghai Biennale are deeply rooted in the unique characteristics of Shanghai, serving as a reflection of its ever-evolving urban landscape. By drawing inspiration from the city's rich history, cultural diversity, and rapid transformation, the Biennale captures the essence of Shanghai and showcases its dynamic energy through contemporary art.

For instance, the 10th edition of the Shanghai Biennale, titled "Social Factory," took inspiration from Shanghai's post-industrial development and the Chinese woodcut movement of the 1920s. This theme explored the intersection of urbanization, industry, and art, showcasing Shanghai as a dynamic hub of social and cultural transformation. The Biennale captured the city's energy and complex social dynamics through artworks and exhibitions, addressing themes such as labor, production, urbanization, and the relationship between individuals and the cityscape. By incorporating elements of Shanghai's historical and contemporary contexts, the Biennale offered a thought-provoking exploration of the city's evolving social fabric and its significance as a global cultural center.

In the 13th edition of the Shanghai Biennale in 2021, the theme of "Bodies of Water" resonated with Shanghai's geographical and cultural context. The city's extensive network of waterways, its proximity to the river and the sea, and its frequent reference as a city made of water provided the inspiration for this theme. "Bodies of Water" explored the profound significance of water in shaping Shanghai's identity, culture, and history. The curatorial approach of the Biennale emphasized the inclusive and open nature associated with water, reflecting on themes of fluidity, connectivity, and the interplay between natural and human environments. Through artistic expressions, the Biennale celebrated the diverse perspectives and experiences related to water, fostering a deeper understanding of Shanghai's unique relationship with this vital element.

As part of the 13th Shanghai Biennale, a captivating video titled "Love Him, Love Her, Love Ta" was commissioned and created by Shanghai-based video artist Yang Fudong. This video garnered widespread attention as it was broadcasted across various platforms. The video featured seven prominent figures from diverse fields, including writer Sun Ganlu, scholar Liu Qing, artist Ding Yi, writer Xu Zhiyuan, K-pop star Jackson Wang, singer Huang Ying, and actress Tan Zhuo. Each participant shared personal stories and connections with Shanghai, reminiscing about significant moments from their past and reflecting on their experiences in the city. The video aimed to underscore the inclusive and diverse nature of the Shanghai Biennale, emphasizing the intimate relationships individuals forge with Shanghai and highlighting the city's role as a vibrant source of inspiration and cultural exchange.

The Shanghai Biennale transcends traditional exhibition boundaries and embraces a dynamic urban narrative. Since 2012, the Biennale has introduced the dedicated "City Project" section, which encompasses a wide range of activities

"藤本壮介展:未来之未来"在五层的SPA空间,2015
Sou Fujimoto: Futures of the Future, 2015
图片提供:上海当代艺术博物馆 *Courtesy Power Station of Art*

"侯赛因·卡拉扬:群岛"展览现场,2021  Hussein Chalayan: Archipelago, 2021
图片提供:上海当代艺术博物馆 *Courtesy Power Station of Art*

such as exhibitions, screenings, field investigations, workshops, and more. These initiatives actively engage with the entire city, extending beyond the walls of the museum. The goal is to foster a deeper connection with the general public and engage them on a spiritual level. One exemplary project that exemplifies this approach is the "51 Personae" initiative, which was introduced during the 11th Shanghai Biennale. This project brought together 51 individuals from diverse professions in the city. Through 51 public activities held at different locations throughout Shanghai, the project offered a dual-level experience of physically going beyond the museum and engaging with the public. This inclusive approach aims to create meaningful and accessible experiences that resonate with a broader audience and encourage active participation in the artistic and cultural landscape of the city.

The Power Station of Art has diligently built its collection, focusing on systematic research on Chinese artists while also showcasing international artists. Architecture holds a continuous and significant place of attention at PSA, highlighting its relationship with people and its ability to speak about humanity and survival. Architects are seen as observers of human society, facing and challenging the problems that arise.

Since 2014, PSA has curated exhibitions featuring renowned architects and studios from around the world, including Sou Fujimoto, Jean Nouvel, John Hejduk, and Heatherwick Studio. Notable exhibitions have explored regional and temporal architectural trends, such as "Action and Recovery: A Trace of Modern Architecture in Taiwan" (2022) and "The Rise of Modernity: The First Generation of Chinese Architects from the University of Pennsylvania." The museum's exceptional exhibition of architecture has earned acclaim within the industry, contributing to the appreciation and understanding of this vital design form.

In addition to its architectural exhibitions, PSA consistently pioneers inventive architectural environments. Back in 2013, the museum introduced the enchanting "SPA Space" on the fifth floor, snugly nestled within a narrow area next to an outdoor riverside terrace. The tranquil atmosphere of this area was meticulously cultivated through the use of pristine white mosaic tiling. A touch of creative humor was injected into the design of two mini "pools," one circular and the other square, providing visitors with a delightful spot to relax and unwind. The nomenclature of the space ingeniously intertwines with the museum's acronym "PSA," aligning harmoniously with the concept of "pools." (A similar infusion of humor is evident in the naming of the "dododo book.")

In 2015, Yona Friedman embarked on the reconstruction of his visionary "Museum of Simple Technology" project, located within the corridor leading to the iconic chimney on the third floor of PSA. Following this, in 2016, Atelier FCJZ introduced an entirely novel space known as the "power station of Design" (psD), serving as PSA's versatile multipurpose design center. The core principle

美术馆通常有着严肃、学术的一面，人们怀着敬仰之心观摩大师作品，仿若一场心灵朝圣；当面对凝聚着高度知识、哲思的作品时，观展也会变成一场高强度的脑力活动。作为上海的城市名片且深度介入中国当代艺术进程的公立美术馆，PSA 在进行宏大叙事的同时，也有着年轻、活泼的一面。

创立于 2014 年的"青年策展人计划"面向全球华人艺术从业者开放，颇有策展人"海选"的意味：计划通过网络公开征集策展方案，获胜者（团队）将得到馆方的支持将梦想落地，策展方案将在馆内得以实现。相比于成熟艺术家的展览，这些来自不同专业和职业背景的策展"新鲜人"带来了"不规矩"和具有刺破性的表达。

年轻的气息也一直在 PSA 的展览画册、衍生品中释放，2023 年春天开张的对对对书店，不出意外地延续了这份活泼——它是"存储"过往展览所有可爱物件的场所。这个斑斓的彩虹颜色、犹如儿童乐园一般的空间，由法国女设计师玛塔莉·克赛特设计。设计师以单轨列车与放射线为灵感，在书店内安放了圆环的列车轨道，并将轨道的色彩与书的色彩、门类对应陈列。书

2023年春天，在PSA一层的沿街空间开张的对对对书店
Unveiled in the spring of 2023, the "dododo book" finds its home on the first floor of PSA, nestled within the vibrant street space

图片提供：上海当代艺术博物馆
*Courtesy Power Station of Art*

of this design centers on the dynamic interaction between the museum and its urban context. The interface between the two is animated by a collection of purposeful shops, strategically positioned to entice visitors to engage with the museum through these enticing "storefronts." This innovative strategy encourages visitors to enter the museum through these inviting entry points and then explore the design center, creating a stimulating pause before venturing deeper into the museum's exhibition halls.

The establishment of psD expanded the museum's coverage in design disciplines beyond architecture to include fashion, jewelry, graphic design, and furniture. Designers often aspire to similar goals as artists, and when their works transition from store shelves to the museum's curatorial system, they take on new symbols and interpretations. Exhibitions like "Hussein Chalayan: Archipelago" (December 2021–February 2022) showcase the transformative power of design, addressing urban and cultural issues through installations, videos, and sounds.

The Power Station of Art provides a serious and academic atmosphere for visitors to observe and appreciate masterpieces, turning a museum visit into an intense mental experience. While engaging in grand narratives, the museum also embodies a youthful and lively spirit. PSA's Emerging Curators Project supports young Chinese curatorial talents, offering them a space to materialize their concepts, gain visibility, and foster professional growth.

The "dododo book", opened in spring 2023, exemplifies the museum's youthful design sensibility. Designed by Mattali Crasset, the bookstore features

店除了有过往及最新展览的出版物、衍生品外,还设有小型展览空间、儿童区、咖啡厅,是让人欢聚、休闲的场所。书店设在 PSA 一层的沿街空间,将美术馆的开放性、日常性进一步扩展,这里也常有建筑、艺术、文化相关的讲座、对谈,向公众免费开放。

上海当代艺术博物馆的演变过程几乎就是一段浓缩的城市变迁史
The evolution of PSA is almost like a condensed history of urban transformation
图片提供:上海当代艺术博物馆 *Courtesy Power Station of Art*

a colorful rainbow palette and circular train tracks that add an interactive and visually stimulating element. With a small exhibition space, a dedicated area for children, and a café, the bookstore extends the museum's openness and accessibility, creating a multifunctional and inviting space for visitors of all ages. The decision to place the bookstore on the first floor of PSA, facing the street, enhances the museum's commitment to openness and everyday engagement. This choice aligns with the architect team's original vision of integrating the museum's positive and accessible attitude into the fabric of urban life.

**参观指南**
◎ 上海市黄浦区苗江路678号
⊕ 周二至周日 *11:00—19:00*（最晚入馆时间 *18:00*）
周一闭馆，国定节假日开放

**Visitor Information**
*◎ 678 Miaojiang Rd., Huangpu District, Shanghai*
*⊕ Tuesday—Sunday 11:00—19:00*
*(last admission at 18:00)*
*Closed on Mondays, open on national holidays*

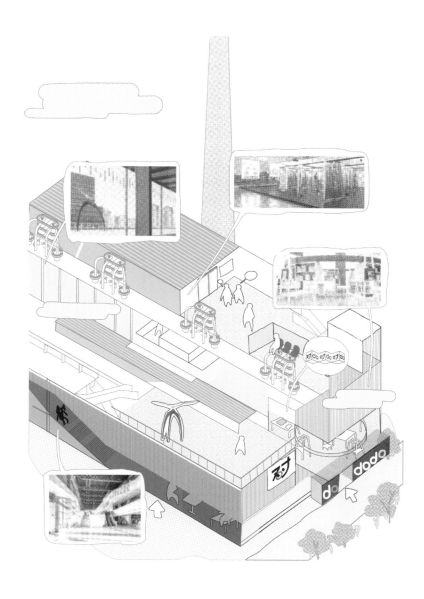

上海复星艺术中心全景　Panorama, Fosun Foundation Shanghai
图片提供：上海复星艺术中心　*Courtesy Fosun Foundation Shanghai*

# 02
# 上海复星艺术中心
# Fosun Foundation Shanghai

建筑师：英国福斯特建筑事务所，赫斯维克工作室 ｜ 开馆时间：2016 年
Architects: Foster + Partners, Heatherwick Studio ｜ Opening Year: 2016

> 与BFC的摩登气质相匹配，该中心的艺术展陈也与都市、都市文化相关。
>
> In line with the modern atmosphere of the BFC, the art exhibitions at the Fosun Foundation Shanghai also encompass urban culture.

外滩万国建筑博览群以南，过了延安东路隧道，城市景观被现代摩登大楼取代——这里是外滩的金融地带。复星艺术中心身处该地段的核心位置，属于上海外滩金融中心（简称BFC）这一综合商业体的一部分。艺术中心是一栋地上四层、地下三层的独立建筑，由英国福斯特建筑事务所和赫斯维克工作室联合设计。建筑外观最大的亮点即三层金色可转动幕帘，如中国古代冠冕，又似西方竖琴，于每天固定的六个时间随着音乐转动，宛如一座"会跳舞的房子"。

复星艺术中心是由复星集团及复星基金会发起并出资建立的非营利机构。复星基金会 2012 年成立后致力于当代艺术的收藏，部分藏品在 BFC 的公共区域长久陈列。比如，宫岛达男的作品《数字空中花园》位于复星艺术中心四层天台；朱利安·奥培的作品《申城漫步》被放置在 BFC 的户外空

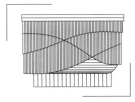

South of the historical building cluster on the Bund and beyond the Yan'an East Road Tunnel, the urban landscape gives way to modern and sleek skyscrapers, forming the Bund Finance Center (BFC), a relatively new financial district in Shanghai. At the heart of this district stands the Fosun Foundation Shanghai, an integral part of the BFC complex. Collaboratively designed by Foster + Partners and Heatherwick Studio, the building comprises four aboveground floors and three below-ground levels. One of its architectural highlights is the captivating facade—an exquisite, rotating golden bamboo-shaped curtain suspended from the third floor. This visually striking feature combines elements from both Eastern and Western aesthetics, evoking the imagery of an ancient Chinese crown and a Western harp. Throughout the day, the curtain gracefully rotates six times, accompanied by music, creating the enchanting illusion of a "dancing house."

The Fosun Foundation Shanghai is a non-profit organization initiated and supported by the Fosun Group and Fosun Foundation. Since its establishment in 2012, the Fosun Foundation has been committed to collecting contemporary art, and a portion of its collection is exhibited in the public areas of BFC. On the rooftop of it, visitors can find the installation "Counter Sky Garden" by Japanese artist Tatsuo Miyajima, while Julian Opie's artwork "Walking in Shanghai" is displayed in the outdoor space of the BFC.

In line with the modern atmosphere of the BFC, the art exhibitions at the Fosun Foundation Shanghai also encompass urban culture. Julian Opie held his inaugural exhibition in China at this venue in 2017. Renowned for his distinctive portraits, Opie depicts figures in a flat, graphic style reminiscent of digital icons or symbols, reflecting the impact of technology on our contemporary lives.

In 2020, the Fosun Foundation Shanghai hosted the first museum-scale exhibition of American artist Alex Katz in China. Born in 1927, Katz is considered one of the most influential artists in the United States during the 20th century. His unique portraits and landscapes have garnered international acclaim. Katz portrays the human form with simplicity and clean lines, flattening both the figures and backgrounds. These portraits are often exceptionally large in size and heavily influenced by the cinema, television, fashion, and billboards that emerged in the 1960s.

In 2021, the Fosun Foundation Shanghai presented "Tadao Ando: Endeavors," a comprehensive documentary on the life and work of the Japanese architect. The exhibition featured architectural models, remakes, sketches, drawings, videos, photographs, and travel logs that documented the architect's journey.

From November 2021 to March 2022, Los Angeles-born artist Alex Israel

间。与 BFC 的摩登气质相匹配，该中心的艺术展陈也与都市文化相关。公共作品遍布全球各大都市的朱利安·奥培在此举办了中国首展，那些极简线条的人形，穿着运动套装，或戴着耳机，是千千万万都市人的灵魂写真；生于1927年的亚历克斯·卡茨创造了现代文化消费场景中的人像典范，他笔下的人像有着电影、广告特写镜头的观感，动人心魄；洛杉矶艺术家亚历克斯·伊斯雷尔作品是彩色和自由的，他把公路、霓虹、海浪带到黄浦江畔。从做城市住宅起步的建筑师安藤忠雄，在80岁生辰之际，在此举办了建筑设计生涯大型回顾展，展厅复刻呈现了其成名作"水之教堂"。

除了艺术大师的展览，复星基金会的艺术收藏展也会在此持续展出。2018年"收藏当代之道：来自复星基金会的影像（2010—2018）"向观众呈现了来自该基金会收藏中的31件影像作品，它们来自26位不同国家和地区、身处不同历史阶段的艺术家和摄影师，媒介丰富，形式多样。

"安藤忠雄：挑战"展览现场，2021　Taodao Ando: Endeavors, 2021
图片提供：上海复星艺术中心　Courtesy Fosun Foundation Shanghai

"亚历克斯·伊斯雷尔:自由公路"展览现场,2021
Alex Israel: Freeway, 2021
图片提供:上海复星艺术中心 *Courtesy Fosun Foundation Shanghai*

"刘建华:形而上器"展览现场,2022
Liu Jianhua: Metaphysical Objects, 2022
图片提供:上海复星艺术中心 *Courtesy Fosun Foundation Shanghai*

"草间弥生：爱的一切终将永恒"展览现场，2019
Yayoi Kusama: All About Love Speaks Forever, 2019
图片提供：上海复星艺术中心 *Courtesy Fosun Foundation Shanghai*

showcased his solo exhibition "Freeway" at the Fosun Foundation Shanghai, bringing his interpretations of iconic Los Angeles motifs such as sunshine, waves, and the sky to the banks of the Huangpu River.

The Fosun Foundation Shanghai also curates ongoing exhibitions featuring its own collection. One notable exhibition is "The Way of Collecting: Images from Fosun Foundation (2010–2018)" held in 2018. This exhibition presented a diverse selection of 31 artworks created by 26 artists and photographers from various countries and historical periods. The showcased works offered the audience a rich and varied range of artistic expressions, providing a captivating exploration of the collection's visual art.

**参观指南**

⦿ 上海市黄浦区中山东二路600号
🕙 周二至周日 10:00—18:00（最后入场时间 17:30）
　周四、周六 10:00—20:00（最后入场时间 19:30）
　周一闭馆（国定节假日除外）

**Visitor Information**

⦿ 600 Zhongshan E-2 Rd., Huangpu District, Shanghai
🕙 Tuesday–Sunday 10:00–18:00 (last admission at 17:30)
　Thursday & Saturday 10:00–20:00 (last admission at 19:30)
　Closed on Mondays (excluding national holidays)

日本艺术家宫岛达男设计的装置艺术作品《数字空中花园》，坐落于上海复星艺术中心的四楼天台
"Counter Sky Garden", an artwork by Japanese artist Tatsuo Miyajima,
is situated on the rooftop of Fosun Foundation Shanghai

图片提供：上海复星艺术中心 *Courtesy Fosun Foundation Shanghai*

外滩1号,东一美术馆位于上海久事国际艺术中心内
Bund One Art Museum is situated in the Jiushi International Art Center at Bund 1
图片提供:东一美术馆 *Courtesy Bund One Art Museum*

# 03
# 东一美术馆
# Bund One Art Museum

开馆时间：2019 年　Opening Year: 2019

作为栖身于国家重点文物保护建筑内的美术馆，东一美术馆的定位是"经典艺术"，展出的是全球知名博物馆的典藏之作。

The Bund One Art Museum, nestled within the walls of the former Asia Building, is a steadfast advocate for world-class art.

从延安东路拐到中山东一路，就到了外滩景区，壮观的外滩万国建筑博览群由此展开。中山东一路的开端，坐落着被誉为"外滩第一楼"的建筑，东一美术馆便栖身于这栋著名建筑内，美术馆的名字也源于这个地址。

这栋建筑建于 1916 年，由当时的马海洋行设计，裕昌泰营造厂施工。楼高 7 层，后顶部四角又增加了 1 层，成为当时外滩天际线"制高点"，加之建筑面积 11 984 平方米的庞大体量，成就了外滩万国建筑群乐章气势恢宏的开篇。不仅宏大，它还有着处处彰显艺术之美的建筑细节：折衷主义风格的外观和巴洛克式的正立面为这份恢宏添了典雅之气；楼内呈"回"字形结构，中间有供人们活动的天井；大楼朝外滩一侧的东面是正门，正门外有 4 根古希腊爱奥尼式的立柱，内门还有小爱奥尼式门柱。该建筑最早因投资兴建的公司得名"麦克倍恩"大楼，1917 年被亚细亚火油公司购得，改名"亚细亚大楼"——这也是它最为人熟知的名字。

45

When we turn left from Yan'an East Road onto Zhongshan E-1 Road, we are immediately welcomed by the awe-inspiring sight of the famous Bund, adorned with a magnificent collection of historical architecture. At the beginning of Zhongshan E-1 Road stands a building that once held the prestigious title of the "Number One Building on the Bund." Today, this iconic structure is the home of our destination: the Bund One Art Museum.

Completed in 1916, the architectural masterpiece that houses the Bund One Art Museum was designed by Moorhead & Halse and skillfully constructed by contractor Yu Chang Tai. Originally standing at seven stories tall, an additional floor was added to the corners in 1939, making it the tallest building on the Bund at that time. This impressive height, combined with its expansive floor area of 11,984 square meters, earned it the widely recognized nickname.

The building's exterior showcases an eclectic architectural style with a touch of baroque influence, adding an elegant charm to its grandeur. Adorning the front facade are stately Ionic columns, while the interior follows the layout of concentric squares, providing panoramic views of the breathtaking scenery on both sides of the Huangpu River.

Initially known as the McBain Building, it was named after the British merchant George McBain, whose company provided the financial support for its construction. In 1917, the building underwent a significant transition when the Royal Dutch Shell's Asiatic Petroleum division acquired a substantial portion of the property and renamed it the Asia Building.

Today, this building has been officially recognized as a Major Historical and Cultural Site protected at the National Level. Its architectural splendor and rich heritage make it an integral part of the Bund's iconic landscape and a cherished treasure of Shanghai's cultural legacy.

Nestled within the walls of this historic building, the Bund One Art Museum is a steadfast advocate for world-class art. With a strong commitment to bridging Shanghai with renowned international art institutions, the museum strives to make art accessible to the local audience.

At the helm of this visionary institution is Executive Director Xie Dingwei, whose lifelong passion for art stems from his artistic lineage. Both of Xie's parents, Xie Zhiliu and Chen Peiqiu, were esteemed calligraphers and painters, immersing him in a world of creativity from an early age. Although Xie initially pursued a different path, earning his M.Eng. in Electrical Engineering from the University of Southern California and working in Silicon Valley for several years, his deep-rooted connection to the art world eventually called him back. Driven by his unwavering vision, Xie transitioned into the realm of museums,

克劳德·莫奈的《日出·印象》在东一美术馆的展览现场，是这幅印象派的开山之作在中国的首展
Claude Monet's "Impression, Sunrise" showcased at Bund One Art Museum, it was the first-ever exhibition of this seminal Impressionist masterpiece in China.
图片提供：东一美术馆 *Courtesy Bund One Art Museum*

作为栖身于国家重点文物保护建筑内的美术馆，东一美术馆的定位是"经典艺术"，展出的是全球知名博物馆的典藏之作。执行馆长谢定伟出身书画世家，为当代著名书画家谢稚柳、陈佩秋夫妇之子。谢定伟在美国南加州大学获得电机工程硕士学位后，在硅谷从事集成电路工作，虽未继承父母衣钵，但自小浸淫的艺术氛围让他深知美育于人的重要性，而"让家乡人民在家门口看到世界最顶级的艺术"也成为美术馆创办的初衷。

上海作为中国现代艺术的重要发源地，引进式的经典艺术展的爆火可以追溯到 20 世纪改革开放之初。1978 年"法国十九世纪农村风景画"展，引发当时全国艺术青年的"朝圣"。艺术家蔡国强回忆说，他首次离开家乡来上海就是为了看该展，"这也是我第一次亲眼看到外国人的原作，才知道莫奈、毕沙罗等艺术家。"1984 年中法建交 20 周年，法国政府带来 25 幅毕加索的真迹，

where he could contribute to the realization of his lifelong dream: making art accessible to all.

Today, as the Executive Director of the Bund One Art Museum, Xie Dingwei combines his rich artistic heritage with his technological expertise, guiding the museum towards new horizons of artistic excellence and cultural enrichment. His unique background and passion for art serve as the driving force behind the museum's dedication to showcasing world-class artworks and fostering meaningful connections between the local community and the global art scene.

Shanghai, as a cultural hub and a forefront of art in China, has played witness to the fervor surrounding imported classic art exhibitions since the early days of the country's reform and opening up. The impact of these exhibitions has been profound, inspiring and attracting young artists from all corners of China. In 1978, an exhibition featuring French 19th-century rustic landscape paintings took place in both Beijing and Shanghai, leaving a lasting impression on the artistic community. Cai Guo-Qiang, a Quanzhou-born artist, fondly recalled that his first visit to Shanghai was specifically to see that exhibition, where he encountered original foreign artworks for the first time and discovered artists like Monet and Pissarro. Another significant exhibition occurred in 1984 during the 20th anniversary of diplomatic relations between China and France, where 25 authentic Picasso works were displayed. Then, in 2004, on the 40th anniversary, Shanghai Art Museum, located near the People's Square at the time, hosted a remarkable exhibition featuring a collection of Impressionist paintings primarily sourced from the Musee d'Orsay in Paris. Xie Dingwei still vividly remembers taking his daughter to see the exhibition, which garnered such immense interest that it took them hours just to enter the venue. These exhibitions have left an indelible mark on the artistic landscape of Shanghai, fueling artistic passion and fostering a deep appreciation for international art.

The Bund One Art Museum, co-founded by Shanghai Tix-Media Co., Ltd. (led by Xie Dingwei, the general manager) and Shanghai XinHua Distribution Group Co., Ltd. in 2019, is a pioneering institution in mainland China. Tix-Media holds the distinction of being the first non-governmental organization dedicated to importing foreign art exhibitions. Their impressive track record includes groundbreaking shows such as "Picasso: Masterpieces from the Musée National Picasso, Paris" (2011), the largest Picasso exhibition in China at the time, and "Master of Impressionism: Claude Monet" (2014), which attracted a record-breaking 400,000 visitors in three months. This record was only surpassed recently by the "Botticelli to Van Gogh: Masterpieces from the National Gallery, London" exhibition at the Shanghai Museum.

The abundance of art exhibitions is a delight for art enthusiasts, but it raises a pressing question: authenticity. With various visual materials and the occasional lack of disclosure regarding replicas or prints in exhibition information, attend-

举办过一次小规模展览；2004 年中法建交 40 周年，中法官方举办了以巴黎奥塞博物馆藏品为主的印象派画展，在位于人民广场旁边的原上海美术馆。当时谢定伟恰好在上海，便带着女儿专程看展，队伍很长，排了两三个小时。

东一美术馆由上海天协文化发展有限公司（简称"天协"）和上海新华发行集团有限公司于 2019 年创立。天协是最早将西方艺术特展引进中国大陆的民间机构，2011 年"毕加索中国大展"是天协办的第一个展，也是当时全国规模最大、展期最长的毕加索作品展。2014 年的"印象派大师·莫奈特展"，3 个月创下 40 万人次的观展纪录。这个纪录直到 2023 年才被上海博物馆的"从波提切利到梵高：英国国家美术馆珍藏展"打破，该展观展数据为 3 个月 42 万人次。

对于公众而言，展览市场的繁荣一方面丰富了人们的文娱生活；另一方面，质量的良莠不齐也增加了民众的"甄别"成本，一些影像制品、仿制品、印刷品展常常在展讯中模糊信息。"是真迹吗？"这成为很多人在看到名家名作珍品类展讯后的第一反应。从 2011 年第一个展开始，谢定伟就开启了与世界级博物馆和艺术机构的合作模式——原汁原味的不仅是作品，还有策展和展陈设计。

"毕加索中国大展"（2011）的合作方是巴黎毕加索博物馆，展陈由该馆馆长亲自设计。"蓬皮杜现代艺术大师展"（2016—2017）展出的是法国巴黎蓬皮杜艺术中心的馆藏，罗朗·乐明担任策展人，他被誉为"法国博物馆界最有创意的人物之一"，2021 年出任蓬皮杜艺术中心主席。2014 年的莫奈特展和东一美术馆成立后的两次莫奈展，合作方均为玛摩丹莫奈博物馆。2021 年东一美术馆与意大利乌菲齐美术馆达成"五年十展"的合作计划，2022 年 9 月"乌菲齐大师自画像"展开幕，展出横跨 16 世纪至 21 世纪的 50 位艺术大师的自画像瑰宝，其中拉斐尔最著名的画作之一——《拉斐尔自画像》首次在上海亮相。2023 年，"波提切利与文艺复兴"展开幕，由乌菲齐美术馆 15 世纪绘画策展人达妮埃拉·帕伦蒂担任策展人，她抵达东一美术馆出席开幕仪式，同时来上海为开展做准备的还有乌菲齐美术馆修复师弗拉维亚·普蒂——乌菲齐美术馆的真迹在出借前都要经过严格检测，以确保这些数百年前的画作能经受得起长途运输的颠簸震动而不被损坏。

除了外方博物馆的策展，天协和东一美术馆还有一位长期合作的展陈设计师玛戈·雷尼西奥。玛戈在法国博物馆界享有盛誉，曾参与卢浮宫、蓬皮杜艺术中心、法国国立图书馆等机构的室内和展示设计，2004 年之后

ees often find themselves asking, "Is this the real thing?"

Authenticity has been a guiding principle for Xie Dingwei right from the beginning, extending beyond the artworks themselves to encompass the curation and scenography of exhibitions. The commitment to authenticity is evident in the exhibitions hosted by Tix-Media and the Bund One Art Museum. For instance, the exhibition "Picasso: Masterpieces from the Musée National Picasso, Paris" in 2011 was curated by the director of the Musée National Picasso, ensuring a faithful representation of Picasso's work. Similarly, "Masterpieces from the Centre Pompidou 1906–1977" from 2016 to 2017 was curated by Laurent Le Bon, a respected figure in France's art world who later became the president of the Centre Pompidou in 2021. The collaborations between Tix-Media and the Musée Marmottan Monet resulted in successful exhibitions such as "Master of Impressionism: Claude Monet" in 2014, as well as subsequent Monet shows, at the Bund One Art Museum.

In 2021, a significant agreement was established between the Bund One Art Museum and the Uffizi Galleries, paving the way for a series of ten exhibitions to be held in Shanghai over the course of five years. The first exhibition created under this agreement, titled "Uffizi Self-Portrait Masterpieces," made its debut in September 2022. This extraordinary exhibition showcased 50 self-portraits by classical masters and contemporary artists, including a captivating piece by Raphael. Building on the success of this initial collaboration, the second exhibition in the series, "Botticelli and the Renaissance," took place from April 28 to August 27, 2023. The selection process of artworks for this exhibition involved careful examination of the panels to determine which paintings could safely travel. Curator Daniela Parenti and restorer Flavia Puoti from the Uffizi Galleries traveled to Shanghai to ensure the meticulous handling of delicate panel paintings. The dedication to preserving the artworks' integrity and the collaborative efforts between the Bund One Art Museum and the Uffizi Galleries continue to bring exceptional exhibitions to art enthusiasts in Shanghai.

In their pursuit of excellence, Tix-Media and the Bund One Art Museum have fostered collaborations with renowned scenography designers, notably including French architect Margo Renisio. Renisio is a prominent figure in the French museum world, having contributed her expertise to interior and scenography design projects for esteemed institutions such as the Louvre, the Centre Pompidou, and the National Library of France. Since 2004, she has been actively involved in China's art scene, lending her talent and experience to create immersive and visually captivating exhibition spaces. Her creative collaboration with Tix-Media and the Bund One Art Museum ensures that the artworks on display are not only presented with authenticity and curatorial excellence but also complemented by thoughtfully crafted and engaging exhibition designs.

Renisio's remarkable design talent was showcased in the 2016–2017 exhibition "Masterpieces from the Centre Pompidou 1906–1977," held in the West Hall

"波提切利与文艺复兴"展览现场,2023　Botticelli and the Renaissance, 2023
图片提供：东一美术馆　*Courtesy Bund One Art Museum*

"从莫奈、博纳尔到马蒂斯：法国现代艺术大展",展览现场,2021
From Monet, Bonnard to Matisse: Masterpieces of Modern French Painting from the Bemberg Foundation, 2021

图片提供：东一美术馆
*Courtesy Bund One Art Museum*

2 of the Shanghai Exhibition Centre. For this exhibition, 72 carefully selected artworks from the Centre Pompidou collection were featured in a unique and captivating manner. The exhibition adopted a timeline approach, with each year from 1906 to 1977 represented by a significant artwork from a renowned artist. Renisio's design took inspiration from the forest of steles found in China, which she encountered during a visit to Xi'an. Drawing on this influence, she conceived an original open scenography that arranged the artworks in a structured and orderly fashion. Her vision included the use of stele-like walls to display the artworks, reminiscent of the calligraphic inscriptions found on the steles. Additionally, she incorporated the colors blue, red, green, and yellow on the back of each stele-like wall as a visual reference to the iconic colorful pipes that adorn the exterior of the Centre Pompidou. This design element served as a subtle connection between the exhibition space in Shanghai and the renowned French institution. The design received high praise from Serge Lasvignes, the president of the Centre Pompidou at the time, recognizing Renisio's innovative and visually striking approach.

开始频繁与中国的美术馆、艺术展览机构合作。由天协引进的"蓬皮杜现代艺术大师展"是玛戈的展陈设计代表作。该展的策展从蓬皮杜艺术中心众多馆藏中挑选了一条叙事线：在 1906 年至 1977 年期间，每一年挑选一位著名艺术家在当年创作的一幅代表作陈列，由此展开了 20 世纪西方艺术史教科书般的叙事。在上海展览中心西二馆的展厅中，玛戈极具创造性地借鉴了中国传统文化中的碑林概念，将 72 位艺术家的 72 件作品如碑林一般错落有致地放置。展厅空旷又有很多大柱子的特点，勾起了玛戈第一次参观西安碑林的回忆，对于古代石碑来说，每一块表面都刻有当时朝廷官员或名人的书法。也正是这个特点启发了其在展厅建造出一片"碑林"的想法，并且在每块碑（墙）上都展出一件艺术作品。此外，为了能与蓬皮杜艺术中心形成呼应，玛戈在每一块碑状作品墙的背面选用了蓝、红、绿、黄四种颜色作为展陈的视觉元素，以呼应蓬皮杜艺术中心建筑立面外大型管道的四种颜色。时任蓬皮杜艺术中心主席塞尔日·拉斯维涅看过此展后表示，上海展的展陈设计比该展此前在日本东京都美术馆的展示现场更让人惊喜。

"波提切利与文艺复兴"展览现场，2023
Botticelli and the Renaissance, 2023
图片提供：东一美术馆 Courtesy Bund One Art Museum

外滩1号建筑外景　Exterior view of Bund 1
图片提供：东一美术馆　Courtesy Bund One Art Museum

# 美术馆的选址很重要

口述 | 谢定伟
| 东一美术馆馆长

我们定位是做经典展览,所谓"经典"就是那些在艺术史上留下痕迹的佳作,包括古典艺术和现代艺术。做经典艺术展览,就要与国际知名博物馆和艺术机构合作,才更有说服力,才能对得起观众。每个大展我们都会有一两年的筹备周期,以应对世界级博物馆的高标准、严要求。

引进毕加索展是一个偶然的机会。2010年年底,一位朋友谈起巴黎的毕加索博物馆要巡展到台北。毕加索博物馆当时正在装修改建,博物馆打算将展品借出做巡展,以此筹集经费。当时博物馆有两套展品,一套在欧美巡展,一套在亚洲巡展,毕加索博物馆馆长希望展览也能到上海。这位朋友问我:你要不要做这个展览?我说可以。于是朋友介绍我与毕加索博物馆馆长见面并谈妥来上海展览,然后是成立天协文化公司,开始筹备展览。

说到办展,开始我最希望是做莫奈展,但毕加索来了也不能错过。做完毕加索展,我就找莫奈展,因为亚洲人都喜欢印象派,而莫奈是印象派中最有名的。于是朋友又介绍我飞去法国拜访玛摩丹莫奈博物馆,协商办莫奈展。西方博物馆出借作品要安排档期,莫奈作品从2013年11月到次年6月有空,刚好可以办两站巡展。第一站在台北,上海是第二站。我们在上海的莫奈展借用了K11艺术中心的场地,业主K11在场地租金及物业管理上给予了极大的支持,使得展览得以成功举办,我们至今感怀于心。2014年春,莫奈展开展,吸引了40万观众。

这两个展给我最大的经验是:展览地点很重要。K11位于淮海路上的黄金地段,交通方便,地铁站出口就是商场B2层,正对着展览售票台;其次,K11周围有办公楼、商业与餐饮,有与展览相匹配的客流。当然,商场办展也

# Location, Location, Location

Narrated by Xie Dingwei
| *Director of the Bund One Art Museum*

Our mission has always been to showcase classic artworks, from the realms of both classical and modern art, that have endured the test of time. To achieve this, we recognize the importance of collaborating with international museums and art institutions, as their prestigious names carry significant weight. However, obtaining loans of artworks from these institutions entails adhering to strict regulations, which is why we invest considerable time and effort, usually spanning one to two years, in preparing for major exhibitions.

The opportunity to host the Picasso exhibition arose by chance. Towards the end of 2010, a friend informed me about an exhibition organized by the Musée National Picasso in Paris that was planning a tour to Taipei. As the museum was undergoing renovations and seeking additional funds, they had two touring exhibitions available, one for Europe and America, and the other for Asia. The director expressed a strong desire to include Shanghai as one of the stops. Intrigued, my friend asked if I would be interested, and I eagerly accepted the offer. A meeting was arranged with the director of the Musée National Picasso, and after fruitful discussions, we established Tix-Media company, marking the beginning of our journey.

When it came to importing exhibitions, my initial choice was to host a Monet show. Impressionism holds great popularity in Asia, and Claude Monet stands as the epitome of this artistic style. However, the opportunity to host a Picasso exhibition was too enticing to resist. Following the success of the Picasso show, I set my sights on realizing a Monet exhibition. A friend recommended a visit to the Musée Marmottan Monet, and we discovered that their collection of Monet works was available for touring from November 2013 to June 2014. This provided us with just enough time to hold a two-stop tour, starting in Taipei and concluding in Shanghai. The exhibition opened in Shanghai during the spring of 2014 and attracted an impressive attendance of up to 400,000 visitors. We secured the K11 Art Mall as the venue, and their support

"莫奈&印象派大师展"展览现场,2021　Monet & Impressionist Masterpieces, 2021
图片提供:东一美术馆　*Courtesy Bund One Art Museum*

in terms of venue rent and property management played a vital role in the success of the show. We are eternally grateful for their assistance.

One invaluable lesson I have learned from these two exhibitions is the significance of location. K11's prime location on Huaihai Road, easily accessible with a subway exit directly facing the ticket counter, ensured a steady stream of visitors. Surrounded by office buildings, shops, and restaurants, it guaranteed a reliable flow of foot traffic. However, hosting exhibitions in shopping malls does have its drawbacks, such as limited space and logistical constraints. Moreover, foreign museums generally prefer traditional venues and are averse to the idea of holding exhibitions in commercial spaces.

In 2019, with the establishment of the Bund One Art Museum, we finally acquired a permanent venue. It is situated within the historic Jiushi International Art Center on the Bund, offering a splendid location. This building, originally constructed in 1916, exudes its own unique charm, although it does come with certain restrictions due to the need to preserve its structural integrity. As a result, it is better suited for displaying easel paintings.

Importing large-scale exhibitions from abroad is not a highly profitable endeavor, and there are instances where we struggle to break even. Nevertheless, our primary goal is to make art more accessible to a wider audience. The subtle impact that art has on one's life is immeasurable, transcending age barriers. I hope that the people of Shanghai can experience world-class art right in their own city.

Currently, the majority of exhibition attendees are young individuals, with women comprising three-quarters of the audience. However, we aim to expand our reach to encompass a broader age range. When the day comes that we witness a significant number of senior citizens among our viewers, we will know that we have succeeded in our mission.

有弊端，比如展厅空间局促、运输通道受限等。此外，外方博物馆对展出场地都有极高要求，一般不愿意去商场办展览。

直到 2019 年有机会成立东一美术馆，我们才有了一个固定的场地。东一美术馆位于外滩的久事国际艺术中心内，地理位置好。这是一栋建于 1916 年的历史建筑，历史建筑有它的吸引力，但也存在保护建筑内部结构的限制，所以改建成展馆后，比较适合做架上绘画展。

引进国外大展，有一些展览甚至无法做到收支平衡，商业盈利是一件难事，我们更看重的是艺术对市民大众的普及。艺术可以在潜移默化中影响人的一生，艺术欣赏可以成为一件从小到老的事，我希望上海人在家门口就能欣赏到世界最顶级的艺术。

现在看展的人群主要是年轻人，年轻女性占了观众人数的四分之三。我们要做的事是将年龄段扩大。等到哪一天中国的展厅里中老年人占了大的比例，那么说明我们展览的业态是发展到了一定的深度。

---

**🚩 03**

**参观指南**

📍 上海市中山东一路1号（上海久事国际艺术中心内）
🕐 每日 10:00—18:00（17:30 停止入场）

**Visitor Information**

📍 1 Zhongshan E-1 Rd., Huangpu Dist., Shanghai
(inside the Jiushi International Art Center)
🕐 Monday–Sunday 10:00—18:00 (last admission at 17:30)

外滩27号，上海久事美术馆位于六层　Bund 27, Shanghai Jiushi Art Museum is located on the sixth floor
图片提供：上海久事美术馆　*Courtesy Shanghai Jiushi Art Museum*

# 04
# 上海久事美术馆
# Shanghai Jiushi Art Museum

开馆时间：2018 年　Opening Year: 2018

三处场馆或独立使用，或联动成为同一展览的共同展场，构成一条流动的外滩观展路线。

The Jiushi museum cluster consists of three venues, each venue in the Jiushi museum cluster has the capability to operate independently or collaborate with the others.

　　久事美术馆群落由三处展馆组成，均位于由久事集团管理的外滩历史建筑内。位于中山东一路 27 号的上海久事美术馆是群落中最主要的展馆，占地面积约为 1 400 平方米。该建筑始建于 1920 年，1922 年竣工，原为英资怡和洋行的办公楼，由马海洋行设计，大楼外观仿英国文艺复兴时期的建筑风格，内部为钢筋混凝土框架结构。久事美术馆群落的另两处展馆分别为久事艺术空间和久事艺术沙龙。前者所在的中山东一路 18 号，原为麦加利银行大楼，被称为外滩 18 号；后者所在的北京东路 230 号原为浙江兴业银行大楼。

　　这三处展馆或独立使用，或联动成为同一展览的共同展场，构成一条流动的外滩观展路线。2023 年"西行画录·东南园墅——建筑师童寯（1900—

The Jiushi museum cluster consists of three venues, all housed in historic buildings on the Bund and managed by Shanghai Jiushi Group. The primary venue, Jiushi Art Museum, occupies the sixth floor of the former Jardine Matheson Building at Bund 27, covering an approximate area of 1,400 square meters. Originally constructed in 1922, this building served as the office of Jardine, Matheson & Co, which was once the largest trading company in China and East Asia. Designed by Moorhead & Halse in a Modern Renaissance style, the structure features a reinforced concrete frame.

The other two venues within the Jiushi museum cluster are Bund 18 Jiushi Art Gallery, located in the former Chartered Bank Building at No. 18 Zhongshan E-1 Road, and Jiushi Art Salon, situated in the former Zhejiang Industrial Bank building at No. 230 Beijing East Road.

Each venue in the Jiushi museum cluster has the capability to operate independently or collaborate with the others. The first large-scale exhibition in 2023, titled "Architect Tung Chuin (1900–1983)'s Special Exhibition," took place across two venues. The exhibition consisted of two parts: "Grand Tour of a Chinese Architect" and "Gardens in South-Eastern China." These parts

外滩18号，久事艺术空间位于二层　Bund 18, Jiushi Art Gallery is located on the second floor
图片提供：上海久事美术馆　*Courtesy Shanghai Jiushi Art Museum*

1983）特展",即采用两场联动的形式,在久事美术馆、久事艺术沙龙呈现两大主题"西行画录"与"东南园墅"。童寯先生是中国建筑界的一代宗师,在上海留下了诸多建筑作品,如大上海大戏院(今大上海电影院)、上海恒利银行(今永利大楼)等,以及久事艺术沙龙所在大楼。

久事美术馆群落各有侧重。久事美术馆把大师真迹带到上海,举办了一系列世界经典艺术作品的展陈,这些大师包括爱德华·蒙克、乔治·莫兰迪、马克·夏加尔、皮埃特罗·阿尼戈尼等。久事艺术空间致力于呈现优秀的海派绘画艺术,探讨海派艺术与中国其他绘画流派,以及中西方艺术间的交流互鉴。久事艺术沙龙更强调展览的未来性和探索性,在这里观众能看到青年艺术家的前卫表达。

北京东路230号,原浙江兴业银行大楼,由华盖建筑师事务所设计,久事艺术沙龙位于一层
Former Zhejiang Industrial Bankbuilding at No. 230 Beijing East Road, designed by the Allied Architects Shanghai (Tung Chuin, Zhao Shen, Chen Zhi).

图片提供:上海久事美术馆
*Courtesy Shanghai Jiushi Art Museum*

"乔治·莫兰迪展"展览现场,上海久事美术馆,2022
Giorgio Morandi Exhibition, Shanghai Jiushi Art Museum, 2022
图片提供:上海久事美术馆
*Courtesy Shanghai Jiushi Art Museum*

"建筑师童寯(1900—1983)特展Ⅱ:东南园墅"展览现场,
久事艺术沙龙,2023
Architect Tung Chuin (1900–1983)'s Special Exhibition Ⅱ: Gardens in South Eastern China, Jiushi Art Salon, 2023
图片提供:上海久事美术馆 *Courtesy Shanghai Jiushi Art Museum*

"卡奥斯狂想曲：西班牙艺术家奥田·圣·米格尔艺术展"
展览现场，久事艺术空间，2021
Kaos Trip: A Color Journey by Okuda San Miguel, Jiushi Art Gallery, 2021
图片提供：上海久事美术馆 *Courtesy Shanghai Jiushi Art Museum*

were presented collaboratively by Jiushi Art Museum and Jiushi Art Salon. Tung Chuin (1900–1983), a renowned master of Chinese architecture, had a significant impact on Shanghai's architectural landscape, including the building that now houses Jiushi Art Salon.

Each venue within the Jiushi museum cluster has its own unique interests and specialties. Jiushi Art Museum has hosted a series of exhibitions featuring internationally acclaimed artists such as Edvard Munch, Georges Braque, Marc Chagall, and Pietro Annigoni. Bund 18 Jiushi Art Gallery typically explores Shanghai regional culture, promoting exchanges between haipai art and other Chinese painting genres, as well as between Chinese and Western art. Jiushi Art Salon, on the other hand, focuses on the future-oriented and exploratory, showcasing avant-garde artworks by emerging young artists.

"巴黎圣母院增强现实感沉浸式展"展览现场，久事艺术空间，2022
Notre-Dame De Paris: L'Exposition Augmemtée, Jiushi Art Gallery, 2022
图片提供：上海久事美术馆 *Courtesy Shanghai Jiushi Art Museum*

## 参观指南

- 上海久事美术馆
  上海黄浦区中山东一路27号6层
- 外滩18号久事艺术空间
  上海黄浦区中山东一路18号2层
- 久事艺术沙龙
  上海黄浦区北京东路230号1层
- 均为周二至周日 10:00—18:00（周一闭馆，17:30停止入场）
  不同展览开放时间会有不同，具体以展览实际开放时间为准

## Visitor Information

- Shanghai Jiushi Art Museum
  6F, 27 Zhongshan E-1 Rd., Huangpu District, Shanghai
- Bund 18 Jiushi Art Gallery
  2F, 18 Zhongshan E-1 Rd., Huangpu District, Shanghai
- Jiushi Art Salon
  1F, 230 Beijing East Rd., Huangpu District, Shanghai
- Applicable to all venues
  Tuesday–Sunday  10:00–18:00 (last admission at 17:30)
  Closed on Mondays
  Please note that opening hours may differ based on exhibitions

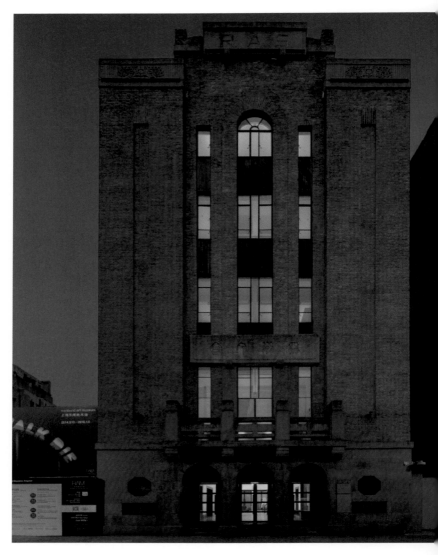

"乌戈·罗迪纳：呼吸行走死亡"展览时的上海外滩美术馆建筑外景，2015
RAM building view during "Ugo Rondinone: Breathe Walk Die" in 2015
图片提供：上海外滩美术馆 *Courtesy Rockbund Art Museum*

# 05
# 上海外滩美术馆
# Rockbund Art Museum

建筑师：乔治·威尔逊（原建筑）；戴卫奇普菲尔德事务所（修复与改造）| 开馆时间：2010 年
Architects: George Leopold Wilson (original design), David Chipperfield Architects (restoration and conversion)
| Opening Year: 2010

> 这个坐落在外滩之源的美术馆，
> 用持久的、系统的、具有稳定质量的项目输出，
> 已成为亚洲当代艺术的重要现场。
>
> Situated at the source of the Bund, RAM has emerged as a significant institution for contemporary Asian art, delivering consistent systematic, stable, and high-quality artistic content.

外滩之于上海，不仅是拥有万国建筑博览群的地标，更是代表着近代上海乃至亚洲城市发展的文化符号。洛克·外滩源百年建筑街区坐落于黄浦江与苏州河的交汇处，由 11 栋历史保护建筑和 6 栋新建筑构成。活跃于 20 世纪二三十年代的知名建筑设计师邬达克、鸿达、李锦沛等都在此留下了他们的传奇作品。2005 年，洛克·外滩源接手项目并于翌年邀请戴卫奇普菲尔德建筑事务所进行 11 栋百年历史建筑修缮，其中包括现为上海外滩美术馆的亚洲文会大楼。

1874 年，皇家亚洲文会（北中国支会）在此地建立了"亚洲文会博物院"（后也被称为"上海博物院"）。这是近代中国第一个向公众开放的博物馆，1886 年门前的马路因此更名为"博物院路"。原建筑后毁于蚁灾，今日所见的亚洲文会大楼完成于 1932 年，由英国建筑师乔治·威尔逊设计，他还设

 The Bund in Shanghai not only reflects the diversity of exotic buildings but also symbolizes the development of modern Shanghai and neighboring Asian cities. Located at the confluence of Suzhou Creek and the Huangpu River, the "Source of the Bund" witnessed architectural activity by notable figures such as László Hudec, C. H. Gonda, and Poy Gum Lee during the 1920s–1930s. In 2005, the "Rockbund project" was launched to revitalize the area, involving the restoration of eleven existing buildings and the construction of six new ones. David Chipperfield Architects was appointed in 2006 to restore, upgrade, and convert the historic buildings, including the former Royal Asiatic Society (RAS) building, now housing the Rockbund Art Museum (RAM).

This building holds significant historical importance. In 1874, the second floor of the structure was opened as a museum by the North China branch of the Royal Asiatic Society of Great Britain and Ireland, commonly known as the Royal Asiatic Society. This museum, later known as the Shanghai Museum, served as China's inaugural public museum. In 1886, the road where the building was situated was renamed Museum Road to commemorate its presence. Unfortunately, the original RAS building had to be demolished due to a severe termite infestation, making way for the current structure designed by British architect George Leopold Wilson of Palmer & Turner in 1932. Wilson's architectural legacy on the Bund remains prominent, with iconic structures like the former HSBC premises, the Shanghai Customs House, and the Cathay Hotel still standing. The RAS building also accommodated an extensive library, boasting an impressive collection of English and Chinese volumes, establishing itself as a preeminent oriental academic research center of its time. Following the Society's closure in 1952, its museum artifacts, mounted specimens, and library collections were dispersed to the Shanghai Museum of Natural History, the Shanghai Museum, and the Shanghai Library.

In the pursuit of preserving the authentic character of the historical building, David Chipperfield made a deliberate choice to maintain the original 1932 design for the building's primary exterior facade during the restoration process. Wilson had skillfully incorporated traditional Chinese decorative elements into the design, resulting in a distinctive architectural style fusion. These elements include three arc-shaped door openings, symmetrical Bagua windows flanking the entrance, cloud patterns, stone lions, and a stone railing on the second-floor balcony. The interior of the Art Deco-style building has been thoughtfully transformed, with newly created flexible spaces that accommodate various presentation concepts. Furthermore, the upper floors from the fourth to the sixth have been interconnected through the addition of a new atrium, enhancing

"宋冬：不知天命"展览时的上海外滩美术馆建筑外景，2017
RAM building view during "Song Dong: I Don't Know the Mandate of Heaven", 2017
图片提供：上海外滩美术馆 *Courtesy Rockbund Art Museum*

蔡国强,《不知如何降下》,"农民达芬奇"展览现场,2010
Cai Guo-Qiang adorned the facade of the building with Chinese characters that read
"Never Learned How to Land" in "Peasant da Vincis", 2010
图片提供:上海外滩美术馆 *Courtesy Rockbund Art Museum*

计了和平饭店、海关大厦、沙逊大厦等9栋建筑，外滩建筑天际线的关键几笔均出自威尔逊之手。亚洲文会大楼曾有丰富的中英文藏书资源，是当时最大的东方学术研究中心。1952年停办后，其积累多年的自然标本、历史文物、艺术藏品和中西文图书分别构成了今天上海自然博物馆、上海博物馆、上海图书馆的典藏基础。

2005年，戴卫·奇普菲尔德受邀负责美术馆建筑改造的设计任务，他保留了原建筑外观，让乔治·威尔逊的设计之魂呈现于世：大楼原本的装饰艺术风格显得典雅精致，同时又因地制宜地汇入中国元素——三个圆弧状的门洞，入口门两侧置有对称的八卦窗，楼体上有着云纹、石狮等图样装饰，二层阳台有石质栏杆。同时，为了满足当代美术馆的功能需求，设计师利用三层通高的新中庭将四至六层连接起来，顶棚上开天窗，重新塑造了内部空间。

上海外滩美术馆开幕于2010年，是洛克·外滩源首个修缮完成、对外开放的项目。这一年，恰逢上海世博会举办，开馆展"蔡国强：农民达芬奇"由享誉国际的艺术家蔡国强策划，在"城市，让生活更美好"的世博大氛围下，该展创意性地提出"农民，让城市更美好"并破除了"艺术是阳春白雪"的刻板印象，让农民群体登上舞台。

从开馆展开始，该馆的项目就体现出对美术馆建筑和其所在街区的极大尊重。这栋建筑的二至四层为展览空间，通常一个展览会用到一整栋楼做叙事，有时还会辐射到五层的咖啡厅、楼顶露台、建筑体之外的"博物院广场"，甚至外墙。在"蔡国强：农民达芬奇"展中，"不知如何降下"这几个字用毛笔书写在一整面墙上；同样是这面墙，2010年末彼得罗·加布里塔·莱斯用灯管组成迷宫，在黑夜绽放。开馆展"蔡国强：农民达芬奇"首次尝试将展览的一部分放在美术馆之外：隔壁的中实大楼的大厅内。此后的"2010曾梵志"展继续这一尝试：在"外滩源"的一座新建教堂，曾梵志以其绘画为素材创作出一组组亮丽的模拟玻璃镶嵌画，将教堂造就成一个光与影的灵境。打破美术馆空间的物理限制，使其能够更好地融入周边、融入社区、融入城市，是上海外滩美术馆一直想要突破的。据馆方介绍，为了首届"RAM Assembles外滩建筑节"，美术馆在邀请戴卫·奇普菲尔德担任首届艺术总监的同时，也携手侯瀚如、李翔宁等策展人团队，共同打造了上海外滩美术馆的崭新入口广场。

近百年前的历史建筑是外滩美术馆的亮点，同时作为中国最早博物馆的意义空间也在当代艺术场域中被延续。群展"时光旅行者"（2012）、"百物

the spatial flow within the building.

The museum served as the inaugural completion within the broader "Rockbund project," and it officially opened its doors in 2010, hosting the exhibition "Peasant Da Vincis" by Chinese artist Cai Guo-Qiang. This event coincided with the Shanghai World Expo 2010, which centered around the theme of "Better City, Better Life." Notably, Cai's exhibition aligned closely with this theme, exploring the idea of "peasants—making a better city, a better life." Curated by Cai himself, the show celebrated the ingenuity of humanity while breaking away from the stereotypical impression of "art is for the elite" and brought the Chinese peasants onto the stage.

Since its inaugural exhibition, the museum has demonstrated a strong commitment to integrating the building with its surroundings. Going beyond the traditional gallery spaces located on the second to fourth floors, the curators have creatively expanded the exhibition areas to include unconventional spaces such as the fifth-floor cafeteria, the rooftop terrace, the "Museum Plaza" (the inner courtyard within the block), and even the exterior facades. For instance, Cai Guo-Qiang, the artist behind the opening exhibition, adorned the side facade of the building with Chinese characters that read "never learned how to land." This same facade later served as the backdrop for Portuguese artist Pedro Cabrita Reis' installation titled "In Here, Out There" towards the end of 2010. Furthermore, some of the contraptions from Cai's exhibition were placed in the nearby National Industrial Bank of China (NIBC) building. Subsequently, artist Zeng Fanzhi followed a similar approach in his show "2010·Zeng Fanzhi." By utilizing the narrow vertical windows of the old Union Church building, Zeng created a mesmerizing array of luminescent virtual stained glass mosaics, transforming the church into a captivating wonderland of light and shadow.

The museum consistently strives to transcend the limitations of its physical space in order to foster stronger integration with its surroundings and the city it occupies. An exciting development on the horizon is the upcoming Bund Architecture Festival called "RAM Assemblies." This event will not only feature David Chipperfield as the inaugural art director but will also involve the participation of curatorial teams led by Hou Hanru and Li Xiangning. One of their ambitious goals is to create a new entrance square for the museum, reimagining the space to further enhance its connection to the city and provide an inviting and engaging experience for visitors. This collaborative effort promises to bring fresh perspectives and innovative ideas to the museum's ongoing commitment to architectural integration and cultural exchange.

The historical building holds a prominent position within the museum, and its rich significance as the former home of China's first public museum is deeply ingrained in the current space. This historical legacy becomes a highlight and is celebrated in various exhibitions. In group shows such as "Time Traveler" in

"乌戈·罗迪纳:呼吸行走死亡"展览现场,2014
Ugo Rondinone: Breathe Walk Die, 2014
图片提供:上海外滩美术馆 *Courtesy Rockbund Art Museum*

2012 and "An Opera for Animals" in 2019, the visual presentation of artworks served as a means to evoke memories of the previous Shanghai Museum, effectively establishing a connection between the past and the present. This deliberate artistic approach not only pays homage to the museum's roots but also sparks a sense of continuity and reflection on the evolving nature of art and culture over time.

Since its establishment, RAM has curated numerous exhibitions and projects featuring extraordinary artists from around the world. Notably, the museum has maintained a consistent and dedicated focus on Asian art. In collaboration with HUGO BOSS, the museum initiated the biennial "HUGO BOSS Asia Art: Award for Emerging Asian Artists" in 2013 (with the 2021 edition being canceled due to the impact of Covid-19). This prestigious award provides a platform for emerging Asian artists to showcase their works to the public. Alongside the exhibition, the museum organizes forums, lectures, performances, and other events that are open to the public, fostering a vibrant cultural exchange. Through this art award, the museum continues its ongoing research and exploration of Asian art.

The exhibition "Tell Me a Story: Locality and Narrative" in 2016 was a notable showcase that drew on artists from diverse Asian regions, sharing eleven captivating stories derived from distinct regional cultures across Asia. The success of

"乌戈·罗迪纳：呼吸行走死亡"展览现场，2014
Ugo Rondinone: Breathe Walk Die, 2014
图片提供：上海外滩美术馆 *Courtesy Rockbund Art Museum*

"菲利普·帕雷诺：共此时"展览现场，2017
Philippe Parreno: Synchronicity, 2017
图片提供：上海外滩美术馆 *Courtesy Rockbund Art Museum*

this exhibition led to its subsequent presentation in Italy two years later, further expanding the reach and impact of the museum's commitment to promoting Asian art and facilitating cross-cultural dialogue.

It seems almost serendipitous that a place once renowned for its dedication to oriental research now immerses itself in the telling of Asian stories through art. By positioning itself as an art institution rooted in the development of Asian art, the Rockbund Art Museum consciously embraces the concept of Asia not merely as a singular continent, but as a vast and diverse collection of regions, nations, and cultures. Situated at the source of the Bund, RAM has emerged as a significant institution for contemporary Asian art, delivering consistent systematic, stable, and high-quality artistic content. Through its exhibitions, programs, and initiatives, RAM has become an essential hub for showcasing and promoting the richness and diversity of Asian art in the contemporary global art scene.

"RAM Highlight 2017：错置"展览现场，2017
RAM Highlight 2017: Displace, 2017
图片提供：上海外滩美术馆 *Courtesy Rockbund Art Museum*

曲"（2019）中，用作品的视觉、展陈方式唤回了人们对上海博物院时代的空间记忆，将这栋建筑的过往与现在接驳了。

上海外滩美术馆一直有着国际性的视野，自从 2010 年开馆后呈现了几十场全球优秀艺术家的展览及项目，从中也能看到一条清晰的亚洲脉络。自 2013 年起，上海外滩美术馆与 HUGO BOSS 联合主办 "HUGO BOSS 亚洲新锐艺术大奖"，两年一届（2021 年因疫情停办）。每一届大奖，候选艺术家的作品都会以展览的形式向公众呈现，展览期间还会举办论坛、讲座、演出等活动（均向公众开放）。以该奖项为契机，馆方以展览、系列讲座、出版等方式，开启了这一持续性的研究方向。"告诉我一个故事：地方性与叙事"展（2016）分享了 11 个来自亚洲不同地区的故事；两年后，这个展览在意大利展出。

可以说是机缘巧合：亚洲文会大楼，这个最初以"亚洲"命名的建筑，继续着亚洲故事的书写。正如其官网上所说的，"源于亚洲持续发展的美术馆，这里的亚洲不仅被视为一个地理意义上的大洲，更是不同区域、国家和文化的集合。"上海外滩美术馆，这个坐落在外滩之源的美术馆，用持久的、系统的、具有稳定质量的项目输出，已成为亚洲当代艺术的重要现场。

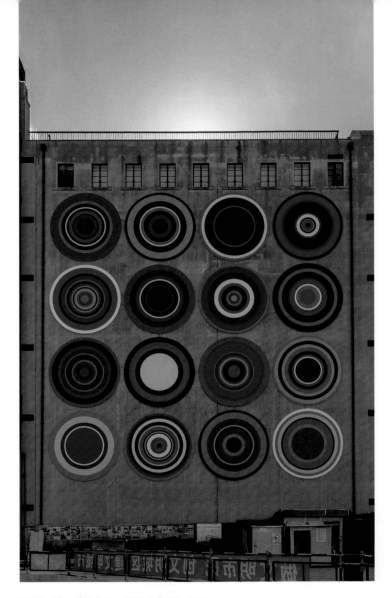

巴尔提·卡尔,《靶心女王》,"轻罪"展览现场,2014
"Target Queen" by Bharti Kher, Misdemeanours, 2014
图片提供:上海外滩美术馆 *Courtesy Rockbund Art Museum*

# 让观众成为美术馆意义建构的合作者

口述 | 刘迎九
| 上海外滩美术馆馆长

上海外滩美术馆从2010年开馆至今（2023年），其发展分为几个阶段：

第一阶段：2010—2012年。我们着力做"原创性美术馆""原创性展览"，希望用展览这种方式来支持艺术家在创作上做出突破。那时合作的艺术家蔡国强、曾梵志、张洹等，都是成熟的艺术家，但是他们在上海外滩美术馆做的项目都是非常新的。

第二阶段：2012—2015年。我们着力扩大美术馆的国际化影响，特别是拉瑞斯·弗洛乔（任期为2012—2022年）馆长上任后，增加与国际艺术家及机构的合作，比如意大利艺术家波拉·彼薇、2013年与法国卢浮宫合作的展览等，拓展了我们美术馆在艺术领域的覆盖范围。相比于第一阶段合作的成熟艺术家，这段时期合作的艺术家正处于职业发展中期，都是非常有实力、有爆发力的。

第三阶段：2015—2018年。这段时间，整个上海的艺术版图发生了巨大变化：2014年西岸艺术与设计博览会创办，次年就已成气候，也出现了一批新的美术馆。而我们美术馆在上海，从空间、资金到人员规模都是不大的，于是调整定位：做一个精品美术馆，不以规模取胜，而是以独特的体验和布展质量来提升观众体验。比如，菲利普·帕雷诺的"共此时"展（2017）。这是一个很"空"的展览，没有固定的作品，起初我们以为观众不一定能接受，甚至都不知道怎么去写导览手册，但最后这居然是我们馆有史以来参观人数最多的展览。展览现场，我们在每一层的空间里都调动了窗帘、灯光、音响……让现场呈现出剧场气质，并且这些东西每时每刻都会根据现场的情况进行转换。

第四阶段：2018年至今。美术馆提出了"海洋性视野"

# Work with Our Audience

Narrated by Liu Yingju
| *Director of the Rockbund Art Museum*

The Rockbund Art Museum (RAM) has undergone several distinct phases since its inauguration in 2010, each with a unique focus and direction. In the initial two years, from 2010 to 2012, RAM positioned itself as an "original art museum" dedicated to showcasing "original exhibitions." The emphasis was on supporting artists in their pursuit of creative breakthroughs. Renowned artists like Cai Guo-Qiang, Zeng Fanzhi, and Zhang Huan collaborated with RAM, presenting groundbreaking works that showcased their artistic innovation within the museum's context.

In 2012, Larys Frogier became the new director, marking a transition for RAM. From 2012 to 2015, RAM intensified collaborations with international artists and institutions. Noteworthy examples include Italian artist Paola Pivi and the exhibition held in partnership with the Louvre in 2013. During this period, RAM primarily engaged with mid-career artists who exhibited remarkable creative prowess.

Between 2015 and 2018, the art scene in Shanghai experienced significant transformations, with the emergence of West Bund Art & Design in 2014 and the proliferation of new galleries. Recognizing the challenges posed by limited resources and competition, RAM strategically shifted its approach to becoming a boutique museum that prioritized the quality of experience and exhibitions over scale. An exemplary exhibition during this phase was French artist Philippe Parreno's "Synchronicity" in 2017, characterized by its unconventional nature. Parreno transformed the museum space through the use of time, space, light, and sound, creating an ever-evolving experience for visitors. The exhibition attracted a record-breaking number of viewers, solidifying RAM's confidence in its new direction.

Post-2018, RAM embraced a self-reflective approach, incorporating an "oceanic vision" and an "institutional critique." Questions arose regarding the boundaries, limitations, and possibilities of an art museum as a cultural and social mechanism. The onset of the Covid-19 pandemic further intensified this reflection and exploration. RAM grappled with how to

及"机构批判"的定位。我们所做的一切都是在反思：美术馆作为一种文化机制、社会机制，它的边界在哪里？局限在哪里？可能性是什么？2019年后遇到了疫情，这给了我们推动思考的契机：当线下的美术馆很难再以原来的方式呈现的时候，我们该如何继续工作？该如何发挥美术馆对于社会、艺术的功能与作用？于是，我们做了很多突破展览外延的努力，比如研究工作、在线杂志等。展览不再只留在美术馆、停留在现场；美术馆也不仅是关于展览的，而展览也不仅是关于一些物件的。

美术馆与观众的关系也在被重新思考：二者不仅是"我提供、你购买"的消费关系，观众应当成为美术馆意义建构的合作者。美术馆生产的是意义——这是一种体验、认识、感悟——而这种意义的实现一定是在合作者的头脑里面，与他自身的已有经验、知识背景相关。美术馆与观众是一种合作关系，美术馆来提供一些东西、一个场所、一些条件，但是观众要付出他相应的行动和努力，来实现艺术的意义。也是基于这个认知，2021年在建筑进行大修时，我们也把一层空间做了重新设计。原先的布局将美术馆和观众的关系切割得太生硬了。新设计用取消柜台、增加流线型布局的方式，希望加强观众的互动感、参与感。

continue its work and fully realize the function and role of art museums within society and art when traditional avenues were no longer viable. Extensive efforts were made to transcend the confines of physical exhibitions, including research work and the establishment of online magazines. Exhibitions expanded beyond gallery spaces, and the museum was recognized as encompassing more than just physical displays of objects.

The relationship between art museums and their audiences also underwent reevaluation, shifting from a provider-consumer dynamic to one of collaboration. Museums are now acknowledged as producers of meaning, facilitating experiences, understanding, and perception. This meaning-making process occurred within the minds of individual viewers, shaped by their unique backgrounds and experiences. In this sense, museums and viewers are engaging in a collaborative dialogue, with the museum providing space, conditions, and stimuli while the viewer actively participates and contributes. This realization prompted a significant redesign of RAM's first-floor space during the building's renovation in 2021. Counters were removed, creating a more fluid and interactive environment, encouraging increased participatory interaction with the audience.

In its evolving journey, RAM has proven itself as a dynamic institution, adapting to changing artistic landscapes and challenging conventional notions of what an art museum can be. Its commitment to pushing boundaries, engaging in critical dialogue, and fostering collaborative experiences continues to shape its role as a vital cultural force in Shanghai and beyond.

**参观指南**

⊚ 上海市黄浦区虎丘路20号
⊙ 每周二至周日 10:00—18:00（最后入馆时间 17:30）
周一闭馆

**Visitor Information**

*⊚ 20 Huqiu Rd., Huangpu District, Shanghai*
*⊙ Tuesday—Sunday 10:00—18:00 (last admission at 17:30)*
*Closed on Mondays*

洛克·外滩源地图　Rockbund map
*本图根据洛克·外滩源提供的地图资料重绘
*Created based on the information provided by Rockbund*

# 滨江大道沿线
# BINJIANG AVENUE

The Lujiazui Financial District, renowned for its iconic "Three Towers," proudly unveiled a national-level art museum in 2021: the Museum of Art Pudong. This esteemed addition has bestowed an artistic and poetic sanctuary upon the very core of this bustling financial district. In conjunction with the Modern Art Museum Shanghai, neighboring galleries, and picturesque public green spaces, it forms an enchanting art belt along the eastern bank of the Huangpu River. This vibrant art realm beckons pedestrians and cyclists, offering them accessible pathways to explore and indulge in the city's artistic treasures.

以"三件套"为地标的陆家嘴金融区，在2021年迎来了国家级的美术馆——浦东美术馆，让这片金融区的中心地带有了艺术的诗意栖息地。由此，其也与艺仓美术馆及附近画廊、公共绿地，勾连成一个黄浦江东岸的滨江艺术带，成为市民徒步、骑行可达的艺术场域。

浦东美术馆建筑外观　Exterior view of the Museum of Art Pudong
图片提供：让·努维尔工作室　*Courtesy Ateliers Jean Nouvel*
摄影：陈灏　*Photo by Chen Hao*

# 06
# 浦东美术馆
# Museum of Art Pudong

建筑师：让·努维尔工作室（设计），同济大学建筑设计研究院（深化）｜ 开馆时间：2021 年
Architects: Ateliers Jean Nouvel (architectural design), TJAD–Tongji University Architectural Design and Research Institute (Group) Co., Ltd. (architects of record) ｜ Opening Year: 2021

浦东，小陆家嘴滨江核心区，
高楼林立间有一片诗意存在。

Nestled amidst towering skyscrapers in the heart of Xiao Lujiazui riverside, Pudong, a hidden gem emerges from the lush foliage and basks in the radiant sunlight—the Museum of Art Pudong.

浦东，小陆家嘴滨江核心区，高楼林立间有一片诗意存在：一座白色建筑掩映在茂盛的树木之中，阳光下显出光泽——这就是浦东美术馆，也是这片寸土寸金的世界级金融区里唯一的公共艺术场所。这座美术馆有着被让·努维尔誉为"诗歌级的地理位置"：西侧与黄浦江及沿江休闲步道相接，与南京路到福州路这一外滩黄金区段隔江相望；向北毗邻上海国际会议中心；向西延伸至黄浦江；向东经由花园广场连接东方明珠塔。

2015 年浦东美术馆确定选址，次年经过四轮评选，让·努维尔工作室所提交的方案胜出。面对 30 米的限高要求，让·努维尔提出了"领地"概念：将后方的公共花园、前场的滨江贯通，都纳入美术馆的空间，53 米长的廊桥自美术馆二层侧门延伸而出，与沿江景观规划衔接，整个美术馆项目所覆盖的面积达到近 1.3 万平方米。让·努维尔希望"浦东美术馆像是沉静融

Nestled amidst towering skyscrapers in the heart of Xiao Lujiazui riverside, Pudong, a hidden gem emerges from the lush foliage and basks in the radiant sunlight—the Museum of Art Pudong. This exclusive public art venue stands as a beacon in the midst of this dazzling financial hub. Renowned architect Jean Nouvel aptly described its geographical location as "epic." The museum is intricately linked to the Huangpu River and the picturesque riverside promenade on its western flank. Across the river, it gazes upon the resplendent stretch of the Bund, spanning from Nanjing Road to Fuzhou Road, radiating an aura of grandeur. To the north, it finds a neighboring companion in the Shanghai International Convention Center, while the Huangpu River extends gracefully on its western side. And to the east, it gracefully connects with the Oriental Pearl Tower via the Garden Square.

The project embarked upon its journey in 2016 when Ateliers Jean Nouvel, the Paris-based studio helmed by the esteemed architect, emerged victorious in a four-round selection competition. Faced with the constraint of a 30-meter height limit, Jean Nouvel ingeniously conceptualized the museum as a "domain," seamlessly blending the public garden at the rear with the waterfront area in front to form an integral part of the museum's space. A magnificent 53-meter-long bridge gracefully extends from the museum's second-floor side entrance, bridging the gap to the Lujiazui riverside walkway platform. Encompassing an impressive expanse of nearly 13,000 square meters, the entire museum project is a testament to Jean Nouvel's vision. The architect envisioned that visitors' journey would commence from the museum's surroundings, stating, "I hope that the Museum of Art Pudong is like a sculpture calmly integrated on the vast land. When people look at it, don't think it's an independent building, but it is a continuation of the collocation of the land, scenery and space sketches, naturally coherent together. This ambiguity is deliberately created by me. I am playing an interactive game with the Huangpu river and the surrounding space," elaborated the architect, a recipient of the prestigious Pritzker Prize.

Nouvel derived the core design concept from the Suprematist art movement, embracing its composition of straight lines, geometric shapes, and color blocks, and translated them into a pure and elegant white box. White serves as the fundamental hue for this remarkable structure. The exterior facade, interior flooring, and outdoor plaza showcase the exquisite Shandong white marble, renowned for its impeccable smoothness and density. The interior space, too, embraces white as its foundational color. The walls are adorned with white latex paint delicately blended with a hint of gray, lending it a more natural and pleasing appearance to the human eye, while providing a suitable backdrop

从展厅望向景观　View towards the garden landscape
图片提供：让·努维尔工作室　*Courtesy Ateliers Jean Nouvel*
摄影：陈灏　*Photo by Chen Hao*

for art exhibitions. The roof's acoustic panels are coated in white, exhibiting a subtle granular texture. This design element satisfies the museum's soundproofing requirements by absorbing sound and minimizing reverberation. The roof's light troughs embody the principles of Suprematism, wherein the architect refines the vibrant color divisions observed in Malevich's paintings into a sophisticated palette of black and white lines.

According to the recollection of the executive manager at the Museum of Art Pudong, an extensive selection of approximately 20 stone samples was meticulously prepared for Jean Nouvel's consideration. The options ranged from polished and matte finishes to flame-treated, pineapple-faced, and lychee-faced textures. Amongst the choices, Mr. Nouvel's discerning eye was drawn to a small, unassuming stone tucked away in a corner, measuring about 30×30 cm. This particular stone featured manually chiseled irregular patterns, imbuing it with a unique texture that could not be replicated by industrial production methods.

Consequently, every stone used throughout the museum, including their size and patterns, is distinctive and not mass-produced. Employing various polishing techniques, these stones acquire distinct textures, carefully tailored to complement different usage scenarios and achieve a seamless integration with the architectural and surrounding landscape.

Incorporating elements inspired by water, the architect ensured that each piece of stone is engraved with patterns reminiscent of raindrops. The slanted lines elegantly capture the drifting motion of rain in the wind. These patterns serve a dual purpose, not only adding aesthetic beauty but also providing effective anti-slip measures for the visitors' safety.

The western facade of the Museum of Art Pudong, facing the majestic Huangpu River, boasts a captivating feature inspired by Marcel Duchamp's renowned artwork, "The Large Glass." It comprises two expansive glass panes, which have become iconic symbols of the museum. These towering panes measure an impressive height of 12 and 6 meters, respectively, and stretch across a width of 55 meters. Remarkably, each pane weighs an astonishing 10 tons.

The glass-fronted facade serves a dual purpose. Firstly, it grants visitors within the museum an unparalleled view of the river, enabling them to immerse themselves in its beauty. Secondly, it seamlessly blends with the surrounding landscape, reflecting the splendid architecture of the Bund on the opposite bank, while simultaneously showcasing the captivating artworks housed within the museum.

These two magnificent glass panes create the Glass Hall, a slender exhibition space adorned with a full-height reflective LED screen. This exceptional hall provides an ideal setting for exhibiting installations, while the LED screen adds a dynamic dimension, capable of showcasing multimedia works, enhancing the visual experience for visitors.

The central exhibition hall, spanning from the basement to the fourth floor,

合在广袤大地上的一方雕塑，人们看过去的时候不要认为这是个独立的建筑，而是和地、景和空间小品搭配延续，自然而然连贯在一起。这种模糊性是我刻意想要营造的，我是在和黄浦江、和周边空间玩一场互动游戏。"

艺术流派中的至上主义被努维尔提炼为核心设计理念。直线、几何形体、平涂色块组合的画面，在建筑大师手中被演绎成纯净、简洁的白盒子。白色是这栋建筑的底色。建筑外立面、室内地面、室外广场的地面均使用了山东白麻——一种极佳的花岗石材料，表面光洁度高，硬度、密度大。室内空间同样以白色作为底色，墙面所用白色乳胶漆被调入了一点灰度，这比纯白看上去更自然，对人眼更友好，也更适合作为艺术品的展陈背景色。屋顶的吸音板喷涂了白色，材质有轻微的颗粒感——这是为了满足美术馆设计的隔音要求，吸音以避免更多的混响。屋顶灯槽的设计也遵循至上主义，马列维奇画作中那些彩色的分割线条被设计师提炼成为黑白线条。

"让·努维尔先生来我们这里选石材样板的时候，桌子上摆了近20种石材，各种样式的，有光面的、哑光面的、火烧面的、菠萝面的、荔枝面的……选定了两三块，起身离开，努维尔走着走着，突然看到墙角放了一块小的石块，差不多30厘米×30厘米的尺寸，他停了下来。"浦东美术馆的工

一层入口大厅全景　　Panoramic view of the ground floor entrance hall
图片提供：让·努维尔工作室　Courtesy Ateliers Jean Nouvel
摄影：陈灏　Photo by Chen Hao

stands as a prominent highlight within the Museum of Art Pudong. Notably, artist Xu Bing contributed a spectacular installation titled "Gravitational Arena," exclusively tailored for this exceptional space. Within the hall, a colossal vortex of characters unfurls, cascading from a height of over 30 meters to the ground. The characters, subjected to the forces of gravity, exhibit captivating tensions and contortions. Adding to the allure, a large mirror on the floor reflects these characters, creating a striking sense of "penetrating" depth between the artwork and the surrounding space.

"Gravitational" alludes to the "wormhole model" formed by the interplay between the installation and its mirrored image, while "Arena" captures the theatrical experience that unfolds as visitors explore the exhibition from various perspectives on different floors. Standing at the base of the installation, visitors are immersed in a sensation akin to being at the center of a grand theater, granting them an up-close view of the intricate details within the artwork. Additionally, "peep windows" are strategically positioned on the second to fourth floors of the central exhibition hall, providing visitors with the opportunity to admire the artwork from diverse angles, further enriching their viewing experience.

The Museum of Art Pudong has forged strategic alliances with esteemed museums of global acclaim, including the revered Tate Modern in the UK, the renowned Metropolitan Museum of Art in New York, and the prestigious National Archaeological Museum of Naples. In 2019, Shanghai Lujiazui Group and Tate established a significant partnership, solidified by a Memorandum of Understanding, as part of the museum's development.

Under this collaboration, Tate generously extends its expertise and knowledge through training initiatives. Moreover, Tate lends its support by presenting an inaugural exhibition derived from its extensive collection, offering a remarkable showcase of artworks to the Chinese audience. Subsequently, two additional exhibitions at the Museum of Art Pudong follow, featuring a carefully curated selection of works sourced from Tate's national collection, thus facilitating cultural exchange and engaging Chinese audiences with internationally acclaimed artworks.

In 2021, the inaugural exhibition titled "Light: Works from Tate's Collection" marked the commencement of this fruitful collaboration between the two institutions. This remarkable project was succeeded by the second exhibition in 2022, aptly named "The Dynamic Eye: Op and Kinetic Art from the Tate Collection." These collaborative endeavors serve as testament to the museum's commitment to delivering extraordinary cultural experiences by uniting the artistic visions and treasures of renowned institutions.

The arrival of esteemed artworks from various corners of the world, including London and Naples, to the Museum of Art Pudong represents more than a mere physical relocation. It serves as the catalyst for the emergence of highly imaginative and professionally curated narratives. "Light: Works from Tate's Collection"

作人员介绍了这段趣闻。这块吸引了让·努维尔目光的石材，上面有着人工凿出来的不规则纹路，这是工业流水线无法出品的质感。浦东美术馆的石材，每一块尺寸、纹路都非流水线出品，它们通过不同的打磨工艺，构建出不同的质感，以配合不同的使用场景，并营造出建筑与周边地景的和谐统一感。作为一座临江建筑，设计师还做了与水相关的设计元素：每一块石材都刻上了如雨滴图案的纹理，斜的直线，是雨在风中飘的线条，同时这些纹路也起到防滑作用。

浦东美术馆西立面由两块"大玻璃"组成，这也是浦东美术馆的标志性景观，设计灵感来自杜尚作品《大玻璃》。"大玻璃"无论从体量到工艺均堪称世界级别，两块"大玻璃"高度分别为12米和6米，宽达55米，单块玻璃重达10吨。"大玻璃"对着黄浦江，观众站在室内可以一览黄浦江景色。同时，"大玻璃"也成为景观的一部分，它既能映射对岸壮观的外滩万国建筑，也反射美术馆内艺术品，室内外双重影像叠印其上。"大玻璃"之内的空间被称为镜厅，一个狭长的展示空间，安置了整面高反光的LED屏幕。镜厅是放置装置艺术的绝佳场所，LED屏幕可用来展示多媒体作品。

除了镜厅，浦东美术馆的中央展厅也是独一无二的世界级展厅。它位于整个建筑的中心，贯穿地下一层至地上四层。艺术家徐冰曾为该展厅量身定做了装置作品《引力剧场》：在这个挑高超过30米的展厅内部，放置了一个从高处垂向地面的巨型文字漩涡，并在展厅底部（即地下一层）设置了巨大镜面。当观众站立于装置底部时，仿佛置身于剧场的正中心，可以清晰看到作品细节——徐冰作品中标志性的独创文字；而地面的镜像使文字嵌入一个巨大的、贯通两个颠倒空间的"虫洞模型"之中，产生强烈的时空穿越感。中央展厅在二到四层都设置了"窥视窗"，方便观众从不同角度观看作品。站在四层观看《引力剧场》，可以看到这些"文字"被拉伸变形，字体由大到小——这是艺术家模拟的引力效果。

浦东美术馆与英国泰特美术馆、纽约大都会艺术博物馆、那不勒斯国家考古博物馆等世界一流的博物馆有着战略合作。早在美术馆开馆前（2019年），上海陆家嘴（集团）有限公司与泰特美术馆理事会就签署了"战略合作谅解备忘录"，约定泰特美术馆为陆家嘴集团建设浦东美术馆提供为期三年的培训和咨询服务，并将在浦东美术馆开馆三年内进行展览合作。2021年的开馆大展"光：泰特美术馆珍藏展"即为双方合作的第一个展览项目；第二个项目是2022年的"动感视界：来自泰特美术馆的欧普与动态艺术馆藏"。这些来

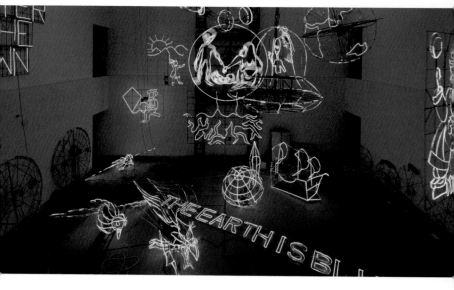

"徐冰：引力剧场"展览现场，2022—2023
Xu Bing: Gravitational Arena, 2022-2023
图片提供：徐冰工作室 Courtesy Xu Bing Studio

"蔡国强：与未知的相遇"展览现场，2021
Cai Guo-Qiang: Encounter with the Unknown, 2021
图片提供：蔡工作室 Courtesy Cai Studio
左下图摄影：顾剑亨 Left below: Photo by Gu Jianheng
右下图摄影：蔡国强 Right below: Photo by Cai Guo-Qiang

embarks on a captivating journey, utilizing light as a unifying element across two centuries of art history. The exhibition delves into the transformative role of light in artistic expression, traversing the mastery of light and shadow by Romantic painters, the impressionists' depiction of light as a subject in itself, the experimental explorations of light in early 20th-century photography, and contemporary immersive light environments where light becomes the very medium of creation. This captivating narrative of art history has garnered tremendous success, captivating almost 400,000 visitors during its exhibition.

The Museum of Art Pudong has been at the forefront of creating significant milestones in the history of Chinese exhibitions. Notably, the inaugural exhibition proudly presented "Ophelia," a treasured masterpiece by British artist Everett Millais, marking its debut in China. This extraordinary inclusion from Tate's collection introduced Chinese audiences to this iconic work for the first time.

Furthermore, the exhibition titled "A World of Beauty: Masterpieces from the National Archaeological Museum of Naples" stands as a monumental achievement, as it represented the first large-scale collaboration between the National Archaeological Museum of Naples and a contemporary art institution in China. The exhibition showcased numerous Italian national treasures, granting them the opportunity to be exhibited overseas for the first time, further enriching the cultural exchange between the two nations.

"The Greats of Six Centuries: Masterpieces from the Museo Nacional Thyssen-Bornemisza" also merits recognition as the Museo Nacional Thyssen-Bornemisza of Madrid presented its collections in China for the first time through this remarkable exhibition. This significant event not only highlighted the museum's diverse masterpieces but also marked the first large-scale overseas exhibition conducted by the museum since its establishment.

Another noteworthy achievement was realized through the exhibition titled "The Shape of Time: Art and Ancestors of Oceania from The Metropolitan Museum of Art." This momentous occasion marked the first-ever exhibition from the esteemed Metropolitan Museum of Art to be showcased in Shanghai, China, signifying a remarkable milestone in cultural exchange and artistic collaboration.

Through these pioneering endeavors, the Museum of Art Pudong has consistently pushed boundaries and set new standards in the Chinese exhibition landscape, providing unique and enriching experiences for visitors while fostering international artistic dialogues.

In addition to its impressive roster of collaborative exhibitions with renowned international institutions, the Museum of Art Pudong also dedicates a series of exhibitions to showcase the works of Chinese contemporary artists. The inaugural exhibition in this series, "Cai Guo-Qiang: Odyssey and Homecoming," was followed by "Xu Bing: Found in Translation" and "Liu Xiangcheng: Lens·Era·People" in 2023, further exemplifying the museum's commitment to promoting

自全球的珍贵藏品来到浦东美术馆，并非只是做了简单的空间挪移，从伦敦到上海，或从那不勒斯到上海，而是在此开启了高度创意和专业性的原创讲述。"光：泰特美术馆珍藏展"用光串联了200年跨度的艺术史，探讨光在艺术创作中的演变：从浪漫主义画家对光影的驾驭、印象派对光线的直接描绘、20世纪初的实验摄影，一直到当代艺术中以光为媒介打造的沉浸式环境。这种充满创意的艺术史讲述大获成功，在展期内共迎来了近40万的参展人流。

浦东美术馆创造了很多中国展览史的第一。比如，开馆大展中的名作《奥菲莉娅》，作为英国泰特美术馆的镇馆之宝，是首次造访中国。展览"绝美之境"是那不勒斯国家考古博物馆与中国现当代艺术机构的首次大规模合作，展览中很多意大利国宝级展品是第一次离开本国参与海外展览。"六百年之巨匠：来自提森‧博内米萨国立博物馆的杰作"是提森‒博内米萨国立博物馆第一次携镇馆之宝来到中国，更是其自建馆以来的首次大型国际展览。"时间的轮廓：大都会艺术博物馆的大洋洲艺术与传承"也是来自大都会艺术博物馆的展览首次登陆中国上海的历史性时刻。

合作展之外，浦东美术馆也会做关于中国当代艺术家的系列呈现。开馆展是"蔡国强：远行与归来"，接着是"徐冰的语言"展，2023年是"刘香成：镜头‧时代‧人"展。作为总建筑面积达4万平方米、设有13个展厅的超大体量美术馆，该馆可同时容纳3~4个大型展览。该馆采用通票制，若参观时间安排妥当，观众一次能观看2~3个展览。

除了看展，浦东美术馆的顶楼天台也值得一去。这是一个360°全景平台。与其他楼宇的屋顶不同，这里看不到任何机电管线及设备这类干扰视线的物件，天台与黄浦江水天一色完美相连，人们在此可看到整个外滩的万国建筑博览群。让视线整洁的秘密是超厚的女儿墙：所有通风管道、排烟管道、机电管线都藏在女儿墙内。另外，地面被架空抬高，在下方走一些小型管线。天台还有一座如透明玻璃盒一般的餐厅，在这里可饱览外滩美景。

and celebrating the talent of Chinese artists on a national and global scale.

With a sprawling floor area of 40,000 square meters and 13 exhibition halls, the museum possesses the remarkable capacity to host three to four major exhibitions concurrently. This expansive space allows for a diverse range of artistic expressions and themes to be showcased simultaneously, providing visitors with a multifaceted and immersive art experience.

To optimize the visitor experience, the museum has implemented a unified pass system, enabling visitors to plan their visit strategically and explore multiple exhibitions within a single visit. By offering this convenient pass system, visitors can immerse themselves in the rich artistic offerings of the museum, ensuring they have the opportunity to engage with at least two to three exhibitions during their visit.

The Museum of Art Pudong's rooftop is a must-visit destination, providing uninterrupted and mesmerizing views of the cityscape. What makes it truly special is the absence of visible mechanical and electrical pipes or equipment, allowing visitors to fully immerse themselves in the captivating surroundings. The secret lies in the clever design of thick parapet walls and elevated ground, effectively concealing ventilation pipes, smoke exhaust pipes, and electrical lines. This meticulous attention to detail ensures a seamless and unobstructed rooftop experience, free from visual distractions.

Adding to the allure, the rooftop features an elegant glass-box-like restaurant. Accessible via elevators numbered 3 and 4 at the western entrance of the museum or the escalators inside, this upscale dining establishment offers a unique opportunity to savor delicious cuisine while marveling at the panoramic city views.

### 参观指南

- 上海市浦东新区滨江大道2777号
- 每日 10:00—21:00（最后入场时间20:00）
  *另每日开放夜场，票价更优惠
  夜场入馆参观时段 17:00—21:00（20:00停止检票）

### Visitor Information

- 2777 Binjiang Avenue, Pudong New Area, Shanghai
- Monday—Sunday  10:00—21:00 (last admission at 20:00)
  For Late night visit (on selected dates only):
  17:00—21:00 (last admission at 20:00)

沿江鸟瞰　Aerial view along the Huangpu River
图片提供：让·努维尔工作室　*Courtesy Ateliers Jean Nouvel*
摄影：章勇　*Photo by Zhang Yong*

艺仓美术馆沿江全景　Panoramic view along the Huangpu River, Modern Art Museum Shanghai
图片提供：大舍建筑设计事务所　*Courtesy Atelier Deshaus*
摄影：田方方　*Photo by Tian Fangfang*

# 07
# 艺仓美术馆
# Modern Art Museum Shanghai

建筑师：大舍建筑设计事务所 | 开馆时间：2016 年
Architects: Atelier Deshaus | Opening Year: 2016

与建筑空间的开放性相呼应，
美术馆推出了艺术、音乐、表演等公众项目，
让这里成为全年无间断的艺术发生现场。

Aligned with the museum's ethos of open architectural spaces, a wide array of public programs encompassing art, music, and performance are offered to ensure a year-round artistic experience.

浦东滨江大道，从塘桥码头到陆家嘴，临江的绿地公园、步道一路贯通，艺仓美术馆便位于这一绿色地带上。该馆前身为老白渡码头煤仓，设计师的改造保留了原有的工业遗构，在满足美术馆展览需求的同时又赋予公共空间以极大的自由度。码头的 8 个煤料斗被保留，"裹进"美术馆内部，成为可供人参观的工业遗迹。煤仓北侧长长的高架运煤通道，被改造成供人行走的高架廊道，连通公共绿地与美术馆。

美术馆主体建筑共 5 层，总面积为 7 065 平方米。除此之外，美术馆所处的滨江大道上，还分布着作为主体场馆延伸的水岸廊桥"艺库"以及翡翠画廊等艺术场馆，共同构造出一个 1.2 公里长的无边界美术馆与艺术江岸舞台。美术馆的主体建筑中，一层为下沉式广场，空间内部保留了原址的煤仓漏斗、铁管，被定位于复合型空间，有餐饮区、艺术商店，以及举办工作坊、

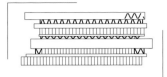

The Modern Art Museum (MAM) Shanghai is situated amidst lush surroundings, spanning from Tangqiao Wharf to Lujiazui along the scenic Pudong Binjiang Avenue. This remarkable institution has been ingeniously repurposed from the historic Lao Bai Du Coal Warehouse, once a bustling coal storage facility. While many of the original structures have been meticulously preserved, including the eight coal hoppers now serving as industrial relics within the museum, they are open for public exploration. The former elevated coal transport passage on the building's north side has been ingeniously transformed into an elevated walkway, seamlessly connecting the museum with the surrounding green public area.

The museum's primary building is a five-story structure covering a vast area of 7,065 square meters. Expanding the museum's influence are two additional venues, namely the waterfront art space known as "Yiku" and the Halcyon Gallery Shanghai. Together, these spaces create an expansive art riverside stage that stretches over 1.2 kilometers.

Upon entering the museum, visitors are greeted by a sunken plaza on the first floor, meticulously designed to preserve the original coal warehouse hoppers and iron pipes. This versatile space accommodates dining areas, an art shop, and serves as a public area for workshops, lectures, and other engaging activities. The second to fourth floors of the museum house galleries, interconnected by a captivating spiral-shaped reinforced concrete staircase. The museum's logo showcases a V-shaped composition created by connecting the coal hoppers, reminiscent of the slender vertical element on the building's facade. This design language is further echoed in the zigzag ramp along the waterfront.

As a prominent cultural landmark in Pudong, the Modern Art Museum has hosted a series of world-class exhibitions since its inauguration. Some notable showcases include "Michelangelo" (December 2016–April 2017), "Giorgio de Chirico & Giorgio Morandi: Rays of Light in Italian Modern Art" (June–September 2017), "Hello My Name is Paul Smith" (October 2017–January 2018), "Bob Dylan: Retrospectrum" (September 2019–May 2020), and "Zaha Hadid Architects: Close Up–Work & Research" (June–August 2021). These exhibitions present the works of esteemed artists and designers, captivating visitors with diverse artistic and creative expressions.

Aligned with the museum's ethos of open architectural spaces, a wide array of public programs encompassing art, music, and performance are offered to ensure a year-round artistic experience. Since 2022, the museum has initiated an ongoing public art project called the "Dock Project," inviting artists from various disciplines to showcase captivating artworks and interactive performances within the museum's public areas. These projects are accessible to the public, providing unique opportunities for engagement and interaction with art.

讲座等公众活动的公共区域；二至四层为展厅，由清水混凝土螺旋形旋转楼梯连通。美术馆的 logo 为煤斗形状连缀的 V 字形，波浪一般，与建筑立面同样 V 字形编织的纤细竖向吊杆相映成趣——这一形式语言也体现在美术馆延伸出的艺仓水岸折形坡道上。

作为浦东的艺术地标，艺仓美术馆开馆后持续推出了一系列世界级大师展，如"米开朗基罗艺术大展""基里科与莫兰迪：意大利现代艺术的光芒""保罗史密斯设计艺术展览""光 / 谱：鲍勃·迪伦艺术大展""扎哈·哈迪德设计回顾展"等。

与建筑空间的开放性相呼应，美术馆推出了艺术、音乐、表演等公众项目，让这里成为全年无间断的艺术发生现场。自 2022 年起，美术馆推出系列公共艺术项目"码头计划"，邀请各领域艺术家在美术馆的天台与廊桥"艺库"等公共区域，展示精彩艺术作品及互动演出，面向公众免费开放。

北侧建筑立面　Northern facade
图片提供：大舍建筑设计事务所　Courtesy Atelier Deshaus
摄影：田方方　Photo by Tian Fangfang

内部空间　Interior view
图片提供：大舍建筑设计事务所　Courtesy Atelier Deshaus
摄影：田方方　Photo by Tian Fangfang

楼梯空间 & "光/谱 鲍勃·迪伦艺术大展"展览现场,2019
Stair space & Retrospectrum Bob Dylan, 2019
图片提供:艺仓美术馆 *Courtesy Modern Art Museum Shanghai*

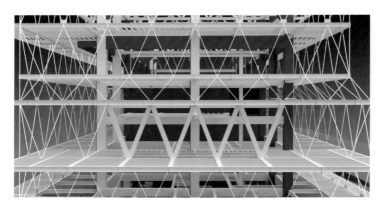

结构效果图　Structure rendering
图片提供：大舍建筑设计事务所　*Courtesy Atelier Deshaus*

### 参观指南

- 上海市浦东新区滨江大道4777号
- 周二至周日 10:00—18:00
  17:30展场停止售票与入场
  周一闭馆

### Visitor Information

- 4777 Binjiang Avenue, Pudong New Area, Shanghai
- Tuesday—Sunday 10:00–18:00 (last admission at 17:30)
  Closed on Mondays

延伸的水岸廊桥"艺库"
"Yiku", the stretching corridor of the waterfront art space
图片提供：大舍建筑设计事务所　*Courtesy Atelier Deshaus*
摄影：田方方　*Photo by Tian Fangfang*

# 西岸美术馆大道
## West Bund Museum Mile

Shanghai West Bund refers to the waterfront area in Xuhui District. It stretches along a shoreline of 11.4 kilometers and encompasses an area of 14.4 square kilometers. In 2010, Shanghai launched the "Comprehensive Development Plan for the Two Banks of the Huangpu River", with the West Bund identified as a key development area. Taking inspiration from successful brownfield revitalization projects such as Hamburg Harbor in Germany and Canary Wharf in the UK, the region underwent the relocation of old industrial facilities and the creation of public open spaces.

Since 2014, a vibrant cultural community has taken shape in the West Bund, featuring the establishment of art spaces like the Long Museum West Bund and the introduction of cultural events such as the West Bund Music Festival, West Bund Architecture and Contemporary Art Biennale, and West Bund Art and Design Fair. From 2019 to 2022, the completion of art museums like Tank Shanghai, West Bund Art Museum, and Start Museum contributed to the formation of the West Bund Museum Mile.

上海西岸位于徐汇区黄浦江畔,岸线长度11.4千米,区域面积14.4平方千米,区域规划总量1340万平方米。2010年,上海市启动"黄浦江两岸综合开发计划",徐汇滨江成为重点建设功能区之一。该区域参考了德国汉堡港、英国金丝雀码头等"棕地"复兴成功经验,完成了旧工业的搬迁以及公共开放空间的打造。

自2014年起,随着龙美术馆(西岸馆)等一批艺术空间,以及上海西岸音乐节、西岸建筑与当代艺术双年展、西岸艺术与设计博览会等文化活动持续亮相,西岸文化群落初具规模。2019年至2022年油罐艺术中心、西岸美术馆、星美术馆等美术馆相继落成,西岸美术馆大道自此形成。

星美术馆全景　Panoramic view of the Start Museum
图片提供：星美术馆　*Courtesy Start Museum*
摄影　*Photo by Moment Studio*

# 08
# 星美术馆
# Start Museum

建筑师：让·努维尔工作室 ｜ 开馆时间：2022 年
Architects: Ateliers Jean Nouvel ｜ Opening Year: 2022

> 以学术的高度去遴选、组织的展览，与消费主义驱动下的"时髦展"有着明显区隔：星美术馆有一颗对待艺术的虔诚之心。
>
> Setting itself apart from consumerism-driven "fashionable shows," Start Museum has established its distinct identity as a sanctuary for art lovers.

　　星美术馆原址是中国近现代历史上第一个海陆转运铁路平台——"日晖港货栈"，始建于清光绪三十三年（1907），后依次更名为日晖港火车站、上海南火车站、南浦火车站。2009 年，为了筹备世博会，上海市政府对滨江沿线进行了整体改造，这座具有百年历史的老火车站才"光荣退休"。如今，这里仍能看到清晰的铁轨，沿着铁轨，可走到龙美术馆。星美术馆旁还停放了一辆货真价实的绿皮火车，以这种形式表达对原场所的记忆。

　　星美术馆是离滨江较近的美术馆，它试图打破美术馆与大众的围墙：最北面的玻璃展厅，24 小时亮灯，几乎是完全透明地置身于公共广场；近江面一侧的墙面由大面积的落地玻璃构成，"人"字形的屋顶庇护着人们，无论天晴下雨都可闲庭信步，即使不进入美术馆也能隔着玻璃看到世界大师的艺术作品。

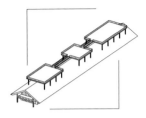

The Start Museum is situated on the grounds of the Rihui Port Warehouse, which served as China's pioneering water-land transport hub and was established in 1907. This historical location was among China's earliest railway stations and underwent several name changes throughout its history, being known consecutively as Rihui Port Railway Station, Shanghai South Railway Station, and Nanpu Railway Station. Ultimately, in 2009, it ceased operations as part of Xuhui's waterfront revitalization initiative in anticipation of the World Expo 2010.

In the present day, clear railway tracks are still visible here. The remnants of the railway have been transformed into a riverside park. Following these tracks all the way, we can reach the Long Museum. On the western side of the Start Museum, a charming verdant-hued train is also parked, beautifully expressing memories of the original place.

Among the various establishments lining the West Bund Museum Mile, the Start Museum holds the closest proximity to the Huangpu River. It strives to foster a stronger connection with the public through its architectural design. The glass exhibition hall located at the northern end remains illuminated throughout the day, creating an impression of boundlessness to those outside. The river-facing facade is adorned with expansive floor-to-ceiling windows, offering captivating views. Additionally, the double-pitched roof provides shelter from unfavorable weather conditions, presenting an ideal opportunity for "window-viewing" the exhibits without entering the museum.

The Start Museum was established by collector and art researcher He Juxing. In 2014, the government extended an invitation to He to create an art museum on the western bank of the Huangpu River that would embody forward-thinking in the realm of international contemporary art. Eventually, the project was entrusted to Ateliers Jean Nouvel (AJN). AJN's design preserved the original layout of the Nanpu railway station's 18-line warehouse, along with its railway tracks and carriages. Adhering to the requirement of maintaining the original building area (approximately 1,777 $m^2$) and height (12 m), AJN fashioned a distinctive exhibition space that resonates with the site's industrial character.

The building boasts a silver aluminum facade and an unobstructed, lofty ceiling, imbuing it with a lightweight and futuristic ambiance. The absence of beams maximizes the interior space, allowing ample natural light to permeate the museum. On sunny afternoons, warm and dazzling light floods the museum's interior. The pitched roof effectively blocks direct sunlight, while scattered skylights enhance the visitor's experience and safeguard the artworks. Jean Nouvel regarded light as a fundamental element—an essential material that could be harnessed—and this philosophy is evident in the design of the Start Museum.

屋顶艺术平台 & 滨江观景平台　Floating art decks & Riverside viewing platform
图片提供：星美术馆　Courtesy Start Museum
摄影　Photo by Moment Studio

The Start Museum represents He's inaugural venture into privately funded art museums, but he is no stranger to the museum scene. Born in 1960, He Juxing immersed himself in the cultural fervor during his university years, evolving into an avid art enthusiast. He delved into contemporary cultural studies and pursued comparative research on avant-garde art. Through cultivating personal relationships with numerous contemporary artists, he began amassing an extensive collection of experimental artworks spanning various genres.

From 2006 to 2010, He Juxing held the position of director at the Yan Huang Art Museum in Beijing. During his tenure, he curated the renowned "Founders of Chinese Art in the 20th Century" series. This period marked an extraordinary era of flourishing Chinese contemporary art. Art-related programs emerged in colleges and universities, with the launch of Art Beijing in 2006, swiftly followed by the proliferation of similar art fairs across the nation. Beijing's 798 Art Zone had already become a cultural landmark, despite the fact that it had only welcomed its first artistic community four years prior, in 2002. Chinese artists began gaining recognition in the international auction market, transitioning from the periphery of society to a more illustrious stage.

In 2010, He Juxing founded the Minsheng Art Museum Shanghai. The inaugural exhibition, "Thirty Years of Chinese Contemporary Art: Painting (1979–2009)," followed by "Chinese Video Art 1988–2011," offered a systematic exploration of the evolution of Chinese contemporary art from a fresh perspective. These exhibitions exemplified the scholarly ambitions of an art museum supported by a private financial institution. He Juxing subsequently spearheaded a series of exhibitions and programs, solidifying the Minsheng Art Museum's reputation as a model for modern museums with a global outlook.

However, He Juxing's involvement with the Minsheng Art Museum came to an end in 2017 when the Redtown International Art Community, where the museum was situated, was demolished. This marked the beginning of a new chapter for He Juxing—the Start Museum—a venture for which he had meticulously prepared for eight years.

The inaugural exhibition, aptly named "START," unveiled a collection of 88 extraordinary artworks contributed by 85 artists exclusively from the museum's own holdings. Each piece within this curated selection shines as a brilliant star, coalescing into a resplendent nebula representing the panorama of global contemporary art from 1960 onwards. This remarkable assemblage serves as a powerful testament to the museum's ambitious mission of shaping art history itself, acting as a resounding manifesto of its creative vision.

He Juxing, as a witness and significant participant in the realm of Chinese contemporary art, has amassed a collection that encapsulates China's cultural landscape over the past three decades. Within this collection, one notable artwork featured in the opening exhibition is Zhang Xiaogang's "Amnesia and

星美术馆由藏家、艺术研究者何炬星先生创办。2014年,何炬星受邀在徐汇西岸筹建一座代表国际当代艺术前沿思考的美术馆。项目邀请了让·努维尔设计,他的设计衔接了过去也拥抱了未来:设计保留了原南浦火车站18号线仓库的平面布局、铁路轨道和车厢,并在保持原有建筑面积(约1 777平方米)和原有高度(12米)的要求下设计展览空间;银色的铝材立面和没有一根横梁的挑高空间赋予建筑轻盈的未来感。无横梁的设计让人惊叹,它不仅最大限度地拓展了室内空间面积,也最大程度地引入了光——每个天晴的午后,日光都会照进场馆。同时,低垂的坡顶遮住了自然光的直射,散射的天光既可以提升观者的感受,又保护了艺术品。"光就是实体,是一种可以运用的基本材料",在让·努维尔看来,光是对于建筑最重要的存在。

这是何炬星创办的第一个私人美术馆,但对于美术馆,他并非新手。出生于20世纪60年代的何炬星大学时代正逢全民化的文艺热潮,那时他便投入对当代文化、当代先锋艺术的研究,和众多同时代艺术家笃信私谊,并

通高展厅　Exhibition space
图片提供:星美术馆　*Courtesy Start Museum*
摄影　*Photo by Moment Studio*

Memory: Man" (2003), personally selected by acclaimed author Yu Hua as the cover image for his novel "Wencheng" (also known as "The Lost City"). Yu Hua and Zhang Xiaogang, both born around the same time as He Juxing in the 1960s, share remarkable parallels in their artistic journeys. They embarked on their creative paths in the 1980s and 1990s, experienced a period of prolific and monumental creativity, and subsequently achieved global recognition after 2000.

Yu Hua was honored with the prestigious Ordre des Arts et des Lettres from France in 2004, while Zhang Xiaogang's "Bloodline Series: Comrade No. 120" fetched a remarkable sum of $979,200 at a 2006 Sotheby's auction of Asian contemporary art in New York, setting a record price for a Chinese contemporary artwork at that time.

A remarkable example within the collection is the monumental artwork titled "Great Migration at the Three Gorges" by Liu Xiaodong. This awe-inspiring piece captures, with astonishing precision, the construction of the Three Gorges Dam—a colossal feat that stands as the largest human-made project in China, if not the world. It evokes memories of the critically acclaimed film "Still Life" by Jia Zhangke, which won the esteemed Golden Lion Award at the 2006 Venice Film Festival. The Mandarin title of the film translates to "The Good Man in Three Gorges." Interestingly, the concept for the film originated from a sketching trip where Jia accompanied Liu Xiaodong to the Three Gorges Reservoir area. Liu Xiaodong, known for his association with China's "sixth generation" filmmakers such as Jia Zhangke, Xiaoshuai, Zhang Yuan, and others, frequently features his friends in his paintings, creating a deep connection between art and cinema.

The exhibition also featured an installation by Maryn Varbanov, a Bulgarian fiber artist who played a significant role in the development of Chinese contemporary art. In 1986, Varbanov established the Maryn Varbanov Tapestry Art Research Institute at the China Academy of Art (formerly known as the Zhejiang Academy of Fine Arts). This groundbreaking institute was the first of its kind in China, dedicated to the creation and instruction of contemporary fiber art. Among Varbanov's early students was Shi Hui, whose work was thoughtfully positioned alongside Varbanov's in the exhibition, showcasing their artistic connection.

Varbanov's wife, Song Huai-Kuei, known affectionately as Madame Song, attained widespread fame as a fashion icon during the 1980s and 1990s. She played a pivotal role in nurturing and promoting China's first generation of models on the international stage. In a unique union, Varbanov and Song Huai-Kuei were married in 1956 with special permission granted by Premier Zhou Enlai. This marked the first mixed marriage since the establishment of the People's Republic of China, signifying a significant milestone in Chinese history.

The opening show also featured an impressive lineup of heavyweight international artists, including Matthew Barney, a prominent American contemporary

开始收藏各类实验性艺术作品。2006—2010年，他出任北京炎黄艺术馆馆长，主持推出"中国艺术二十世纪奠基人系列"。那段时期正是中国当代艺术迅猛发展期，在学科建设、展览活动和市场维度均有体现：全国各地高校纷纷开设艺术专业；2006年，艺术北京博览会创办，此后艺博会模式又辐射到更多城市；这一年，北京798已成为文艺地标，而仅仅是四年前也就是2002年，这里才迎来首批艺术家入驻；中国艺术家在国际拍卖市场上屡创佳绩，完成了艺术品货币化的积累，艺术家群体也从边缘化走向光鲜舞台。

北京炎黄艺术馆之后，何炬星于2010年在上海创办民生现代美术馆。开馆展"中国当代艺术三十年历程·绘画篇1979—2009"及其后的"中国影像艺术：1988—2011"，以全新视角系统梳理了中国当代艺术的发展历程，展现了一座美术馆的学术雄心。其后民生美术馆的一系列展览和文化项目，用具有开拓性的视野在国内建构了一个全新的国际化现代美术馆样本。随着2017年淮海西路"红坊"地块的拆除，何炬星的这段美术馆旅途也告一段落。2022年年末星美术馆开幕，何炬星为此足足准备了8年。

星美术馆的开馆展"开启START"，由全球的85名（组）艺术家、88件作品组成。这88件作品每一件都是一颗璀璨的星，构成了1960年后世界当代艺术的壮丽星云。这些展品均为星美术馆藏品，以管窥豹，可以看出星美术馆建构"美术史"的野心。

作为中国当代艺术进程的见证者和重要参与者，何炬星也用藏品勾勒了国内三十年文化圈的生态图。"开启START"中展出的张晓刚作品《失忆与记忆：男人》（2003），被作家余华选为小说《文城》的封面图。余华、张晓刚、何炬星均为20世纪60年代左右出生，余华和张晓刚在20世纪八九十年代开始创作并迅速进入创作的黄金期，在2000年后迎来世界级的荣誉：2004年余华被授予法兰西文学和艺术骑士勋章，2006年纽约苏富比拍卖行举行的"亚洲当代艺术"专场上，张晓刚的作品《血缘：同志第一百二十号》以97.92万美元成交，创造了当时中国当代艺术成交价的最好成绩。一层展厅中刘小东巨幅画作《三峡大移民》（2003），记录下中国乃至全球历史上最庞大的人造工程——三峡大坝建设。而2006年荣获威尼斯电影节金狮奖的影片《三峡好人》，正萌生于贾樟柯陪刘小东到三峡库区写生时。刘小东与贾樟柯、王小帅、张元这批中国第六代导演相识于20世纪八九十年代，是从学生时代就一起相伴、成长的挚友，这些好友也被刘小东画进作品。一层展厅有一件万曼的装置作品，这位保加利亚籍艺术家深度介入了中国当

artist and film director. The exhibition showcased works by Anselm Kiefer, a renowned German Neo-Expressionist artist, and Nam June Paik, widely recognized as "the father of video art." This remarkable gathering of artistic talent provided art enthusiasts with a feast for the senses, offering a unique opportunity to appreciate extraordinary works without the need to travel abroad.

The Start Museum, through its meticulously curated and thoughtfully organized exhibitions, has demonstrated a sincere dedication to the realm of art. Setting itself apart from consumerism-driven "fashionable shows," Start Museum has established its distinct identity as a sanctuary for art lovers. The exhibitions exemplify the museum's commitment to presenting exceptional art and creating an immersive and enriching experience for its visitors.

代艺术进程。1986年,他在中国美院(时称浙江美术学院)创立"万曼壁挂艺术研究所",这是中国第一个从事当代纤维艺术创作与教学的机构。施慧是万曼的第一批学生,在"开启START"展览中,她的作品被放置在万曼作品旁。万曼的太太宋怀桂是中国第一代模特的发掘者与培养者,他俩于1956年结婚,是由周恩来总理特批的新中国第一桩涉外婚姻。

除了中国艺术家,星美术馆的开馆展"开启START"中还有诸多世界级大师,如先锋电影领域大师马修·巴尼、德国新表现主义大师安塞姆·基弗、影像之父白南准等。对于大众而言,这是一次无需出国门就能密集领略大师作品的盛宴。以学术的高度去遴选、组织的展览,与消费主义驱动下的"时髦展"有着明显区隔:星美术馆有一颗对待艺术的虔诚之心。

沿江鸟瞰　Aerial view along the Huangpu River
图片提供:让·努维尔工作室　*Courtesy Ateliers Jean Nouvel*
摄影　*Photo by Moment Studio*

星美术馆前的绿皮火车是原址场域（日晖港货栈）的表征，也是美术馆的儿童美育空间
The verdant-hued train in front of the Start Museum symbolizes the Rihui Port Railway Station, the original site, and also serves as the museum's children's art education space
图片提供：星美术馆 *Courtesy Start Museum* 摄影 *Photo by Moment Studio*

# 重启新的艺术世界

口述 | 何炬星
| 星美术馆馆长

我和让·努维尔先生第一次见面的地方在他位于巴黎的寓所。我首先讲了五点，希望他能理解做星美术馆这座建筑不仅是个人的邀请，也代表了我们对这座建筑的未来判断。第一，这里是中国近代第一座水陆转运站，代表中国末代皇朝也曾经有过的工业时代梦想；第二，战争时期，这里被外国炮舰占领，被轰炸威胁，见证了动荡与坚韧；第三，这里曾是水路与陆路转运相接，货物和出行的人们走向四方的起点；第四，有沿岸面江 100 多米的建筑宽度，自然与人文环境最适宜做美术馆；第五，我有近 40 年的文化研究和涉及 20 世纪 60 年代以来的国际当代艺术私人收藏。以上五点都值得我们共同打造一个具有国际视野和影响力的美术馆建筑。让·努维尔的建筑作品不仅具有超强的个人情感，他对事情的判断也同样充满智慧和文化情怀。我们首次见面他便欣然接受我的设计邀请。接着，我陪他去尼斯度过周末时间，开始一起讨论建筑构想。他一边跟我说话，一边兴奋地亲手画起草图，也许建筑的雏形就是由此诞生的。

在星美术馆筹备的八年间，完成美术馆建筑的改建是基础。但我更难决择的是"要重建一个什么样的美术馆"。我在 2008 年创建民生美术馆时，思路很清晰：中国当代艺术发展三十年了，需要一座可以对话国际的美术馆。那些年，我和艺术家朋友们都还年轻，有共同的期待和梦想。但现在，国内的当代美术馆在我筹备建筑的八年间，雨后春笋般地纷纷建立起来。不仅如此，当代艺术环境也发生了巨大变化，艺术家似乎不再那么需要一个过时的美术馆了。年轻人的小宇宙和分子革命已经成功，他们对于"小我"之间的内卷已经驾轻就熟。而如繁花一样的美术馆时代我似乎也适

# Re-starting a New Art World

Narrated by He Juxing
| *Director of Start Museum*

My first encounter with Mr. Jean Nouvel took place in his Paris apartment. I needed him to understand that the museum project was not just a personal commission, but a manifestation of my vision for the future of this building, and my profound aspirations for a new art world. I presented him with five essential pieces of background information. Firstly, the original structure was one of the earliest railway stations in China, constructed in 1907 as a glimmer of industrial hope during the last imperial dynasty. Secondly, this place endured occupation by foreign gunboats and wartime shelling, witnessing historical turbulence and displaying resilience. Thirdly, it served as a pivotal water-land transport hub, where people and goods embarked on journeys in all directions. Fourthly, boasting over 100 meters along the riverfront, its natural and cultural environment was highly suitable for a museum. Lastly, having dedicated nearly four decades to cultural research, I had amassed an extensive collection of international contemporary art since the 1960s. I believed these ingredients were sufficient to create an art museum architecture with a global vision and impact. Mr. Jean Nouvel's architectural works bore strong personal influences, reflecting remarkable intelligence and cultural awareness. To my delight, he accepted the commission. We later traveled to Nice for the weekend, engaging in enthusiastic discussions and sketching out initial design ideas.

The preparation for the Start Museum took eight years to complete. The renovation of the original building was of utmost importance, serving as the foundation for everything that followed. However, another question plagued my thoughts: What kind of art museum did I truly want to create? When I embarked on the creation of the Minsheng Art Museum in 2008, my intentions were clear: Chinese contemporary art had been developing for three decades, and it was high time to establish a museum that could engage with international audiences. At that time, my artist friends and I were youthful and shared ambitious aspirations. The Start

"开启START"展览现场,2022  START, 2022
图片提供:星美术馆 Courtesy Start Museum
摄影 Photo by Moment Studio

上图:"开启START"展览现场,2022　Above: START, 2022
下图:马修·巴尼作品《狄安娜》在"开启START"展览现场,2022
Below: "Diana" by Matthew Barney was exhibited at "START", 2022

图片提供:星美术馆　*Courtesy Start Museum*　摄影　*Photo by Moment Studio*

应不了，但我仍然固执地相信会出现一个新的构建。这当然毫无疑问会使我们成为少数派甚至孤独者。而且，星美术馆的地理位置虽说在西岸艺术带的起点，但同时周边尚在开发中，缺少成熟的艺术氛围。所以我在兴奋的同时也有巨大压力。上海是全世界少有的具有先锋意识和市民高素养的城市，要建一个什么样的当代美术馆才可以与之匹配呢？这些问题无时不在困扰着我。

现在，星美术馆已经开馆了，我就希望在上海这样一座深藏巨大文化能量的城市里，它可以作为一个深度研究的当代艺术国际平台，从这点来说，我认为其是具有一定优势的。同时，我选择20世纪60年代以来的国际当代艺术作为星美术馆基本的研究实验范畴，明确了我的命题和目标，我有信心在这个目标上有所作为。从某种程度上来说，我是在做一件全新的事情。因为一般意义上的美术馆有很多现成的范例和专业摆在那里，经验复制各取所需。我给自己出了一套从无到有的命题，难度和生涩感可想而知，但在上海这样的城市，我相信一定可以聚拢更多同频共振的人，甚至可能在年轻人和大众层面引起更多的争论和好奇心。

当然，要做一个全新的美术馆，是要排除很多困难的，包括建立星美术馆所重塑的艺术历史视野，我们理解的艺术、艺术思想和艺术家，以及我们所展开的新的工作方法，等等。我们既要学习现有体系、现成语言和方法之间的关系，又要跳出来完成自己布下的实验命题。总之，我们也刚刚开始，在理想展开之时，还需要更多志同道合的文化理想者共赴时艰、迎接挑战。

Museum, however, was different. Over the course of eight years, circumstances had changed drastically. Contemporary art museums were springing up at an astonishing pace. The artistic landscape was undergoing tremendous shifts, rendering traditional art museums seemingly irrelevant. The younger generation had concluded their molecular revolutions and had become accustomed to the involution of their "small selves." These changes were overwhelming, to say the least. Nevertheless, I maintained a firm belief that the Start Museum would be unlike any other. This would undoubtedly position it as a minority, perhaps even a solitary endeavor. Furthermore, despite its unique location as the starting point of the West Bund Museum Mile, the surrounding area was still under development, lacking a pronounced artistic vibe. So, while I was undoubtedly excited, I also felt immense pressure.

Finally, after much anticipation, the museum is open. Nestled within a city fueled by cultural energy, I hope it can serve as an international platform for in-depth research on contemporary art. I believe it possesses distinct advantages in this regard. I have chosen international contemporary art since the 1960s as the museum's primary research field, an objective in which I am confident we can make significant strides. In many ways, I am embarking on uncharted territory. While general art museums have countless precedents and specialized fields from which they can draw inspiration, I have chosen to tread a rugged and unconventional path, imposing upon ourselves an unprecedented assignment. Yet, in a city like Shanghai, I am steadfast in my belief that we will attract kindred spirits, sparking lively debates and curiosity among both the youth and the general public.

Undoubtedly, the journey to establish a brand-new art museum comes with numerous challenges. We must navigate the reshaped art historical perspective, redefine art, artistic ideas, and artists themselves, and develop new methodologies, among other considerations. We need to strike a delicate balance between existing systems, languages, and methods while daring to think outside the box and complete the experimental assignment we have set for ourselves. In essence, this is just the beginning. We will require the support of like-minded individuals who share our ideals, working together to overcome the challenges that lie ahead.

**08 参观指南**

- 上海市徐汇区瑞宁路111号
- 周二至周五 10:00—17:30，周一闭馆
  周末及公共节假日 10:00—18:30

### Visitor Information

- 111 Ruining Rd., Xuhui District, Shanghai
- Tuesday—Friday 10:00—17:30
  Closed on Mondays
  Weekends & national holidays 10:00—18:30

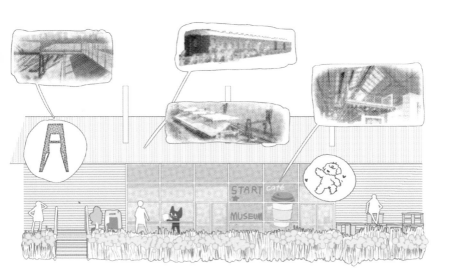

龙美术馆主入口　Main entrance of the Long Museum West Bund
图片提供：大舍建筑设计事务所　*Courtesy Atelier Deshaus*
摄影：苏圣亮　*Photo by Su Shengliang*

# 09
# 龙美术馆(西岸馆)
# Long Museum West Bund

建筑师:大舍建筑设计事务所 | 开馆时间:2014 年
Architects: Atelier Deshaus | Opening Year: 2014

> 对于民众而言,这座美术馆不仅仅是美术馆。邻近龙腾大道的广场种植的160余株樱花树每年初春都会如约绽放,成为浪漫花海。
>
> The Long Museum West Bund transcends being a mere destination for art enthusiasts; it serves as a dynamic catalyst within the community. During early spring, over 160 cherry trees transform the square near Longteng Avenue into a captivating sea of blossoms.

瑞宁路与龙腾大道的路口,一片开阔的广场,地面上镶嵌着几条铁轨,宛如落雪大地上的黑色车辙。沿着铁轨走约80米,就是龙美术馆(西岸馆)(简称"龙美术馆")。清水混凝土、张拉金属网、玻璃幕墙共同构建的清冷灰色与一旁由多个料斗组成的运煤平台旧址,浑然一体。从广场上望去,煤漏斗两侧的建筑,有着圆润曲线的T字形——如在天空上撑开的两把伞。

龙美术馆的原址北票码头,有着近百年历史,曾为上海港装卸煤炭的专用码头,是20世纪末华东地区的能源中心。大舍建筑设计事务所(简称"大舍")进行设计时,保留了铁轨、煤漏斗这些工业时代的遗迹。美术馆主体建筑以独特的"伞拱"结构为建构特征,地上地下共分为四层。地上一层、二层大尺度出挑的拱形空间由质地细腻的清水混凝土浇筑而成,与原北票码头构筑物煤漏斗改造而成的时尚空间"斗廊"形成视觉呼应。这是大舍的第一个美术馆作品,一战成名:不仅开启了中国美术馆空间的全新想象力,也

A pair of railway tracks traverse the expansive square at the intersection of Ruining Road and Longteng Avenue in Xuhui district, resembling imprints in the snow made by passing cars. About 80 meters away from the tracks, the Long Museum West Bund stands tall. It boasts a sleek appearance with its concrete surface displaying an as-cast finish, complemented by tensile wire meshes and glass curtain walls. The museum's cool gray and stony appearance blends seamlessly with the adjacent coal-hopper unloading bridge, a testament to the area's industrial history. When viewed from the square, the buildings flanking the unloading bridge showcase elegant T-shaped curves, reminiscent of two sunlit umbrellas stretched out.

Prior to its magnificent transformation, the site was home to the Beipiao Wharf, which had stood for nearly a century. Dedicated solely to coal transportation, it served as the vital hub for energy distribution in East China during the late 20th century. Atelier Deshaus has preserved remnants from this industrial era, such as the tracks and the coal-hopper unloading bridge. The museum's main building, characterized by its unique "vault-umbrella" structure, comprises four floors, two above ground and two below. The upper floors feature a striking cantilevered arched space, constructed with intricately textured as-cast-finish concrete, seamlessly blending with the modernized "Hopper Corridor," an innovative space that repurposes the former coal-hopper unloading bridge.

As Atelier Deshau's inaugural museum project, the Long Museum West Bund, not only solidified the architect's reputation but also presented a fresh perspective on museum design, igniting the development of Xuhui's own "museum mile."

The Long Museum, established by art collectors Liu Yiqian and Wang Wei, is a private art museum renowned for its diverse collection. Liu specializes in traditional Chinese art and miscellaneous items, while Wang has a preference for Chinese revolutionary art and international contemporary art. The museum has played a pivotal role in introducing many renowned artists to China for the first time. Notable artists showcased at the Long Museum include Olafur Eliasson, an Icelandic-Danish artist known for his sculptural and large-scale installation art utilizing elemental materials; Chiharu Shiota, whose thread installations fill entire spaces to express the essence and intangible aspects of life; James Turrell, a prominent artist associated with the Southern California Light and Space movement of the 1960s and 70s, who employs light to create immersive spatial experiences; and Mark Bradford, whose artworks feature gestural palimpsests of paint, canvas, paper, rope, and everyday materials, challenging conventional perceptions of two-dimensional paintings.

In contrast to the conventional white boxes, the "vault-umbrella" structures present

奏响了徐汇滨江艺术带的创新乐章。

龙美术馆由藏家刘益谦、王薇夫妇创办，他们的收藏谱系能清晰地在展览上体现——从刘益谦专攻的中国传统艺术、杂件到王薇专注的"红色经典"和世界现当代艺术。很多当今活跃在世界舞台的顶级艺术家在龙美术馆完成了在中国的首次亮相。比如奥拉维尔·埃利亚松，这是一位擅长制造自然景观的北欧艺术家；盐田千春，她用毛细血管般密密麻麻的红线占满空间，贲张出生命的活力与迷乱；詹姆斯·特瑞尔，美国"南加州光和空间运动"先驱人物，用光来创造空间，带人进入灵性的冥想世界；出身劳工家庭的美国黑人艺术家马克·布拉德福特，用颜料、画布、纸张、绳索和日常材料不断重叠，颠覆了人们对平面绘画的认知。

与寻常艺术空间那种平整光亮的"白盒子"不同，这个超高的有着洞穴感的空间挑战并刺激着艺术家，为了"征服"它，艺术家往往要花费一两年的时间去准备，去创作更大、更有力量的作品。可以说，展厅空间激发了艺术家新的创作。生于20世纪60年代的上海艺术家丁乙，带有工业设计感的"十字"符号是他的标志。为了让作品有足够的力量去抗衡这个冷峻的清水混

这个超高的有着洞穴感的空间挑战并刺激着艺术家
The "vault-umbrella" structures in upper floors create a stimulating environment for artists
图片提供：大舍建筑设计事务所　*Courtesy Atelier Deshaus*
摄影：夏至　*Photo by Xia Zhi*

in the upper floors of the museum establish a striking cantilevered and cavernous environment that stimulates artists. It is not uncommon for artists to invest more time than usual in preparing for exhibitions here, aiming to create larger or more striking artworks that can effectively resonate with the expansive space.

In the exhibition "What's Left to Appear" held in 2015, Ding Yi, a renowned Shanghai-based artist recognized for his distinctive crosses and grids, showcased a series of new works on plywood. Intentionally created on a grand scale, keeping in mind the expansive central space of the Long Museum, the artist employed thick layers of paint reminiscent of lacquer, which he meticulously scored through to unveil vibrant layers of color. As always, his trademark grids provided a structured framework for the compositions. The size of each artwork, standing at nearly 5 meters tall, necessitated the use of an electric crane during installation.

In 2013, Olafur Eliasson visited the site of the Long Museum before its completion and was deeply inspired by the cavernous atmosphere of the space. Fast forward to 2016, he welcomed visitors to his exhibition titled "Nothingness is not nothing at all," presenting a captivating natural utopia crafted through a series of new artworks. Eliasson's creative vision harmonized seamlessly with the architectural elements of the museum, which combined rectangular rooms with arched ceilings. Within this context, the artist employed fundamental geometric principles, incorporating circles, spheres, cubes, and pyramids into his artworks.

The expansive and monumental space offered by the Long Museum allows for exhibitions and artworks of truly exceptional magnitude. In 2021, Chiharu Shiota presented her largest-ever solo exhibition, "Chiharu Shiota: The Soul Trembles," within these walls. In 2018, the museum showcased the awe-inspiring 10-meter high giant spider sculpture titled "Maman" by French-American artist Louise Bourgeois in its lobby. Bourgeois, born in 1911 and widely regarded as one of the most significant artists of the 20th and 21st centuries, had a prolific career that spanned almost every artistic trend in the latter half of the 20th century. Her iconic spider sculptures, symbolic representations of her mother, can be found in public spaces across major cities worldwide. The aforementioned spider sculpture was part of her first large-scale museum exhibition in China titled "The Eternal Thread." It originated from the Tate Modern collection and was so massive that it could only be installed outdoors or within a building of industrial scale.

The Long Museum serves as a prominent platform for contemporary Chinese artists, including Zhang Xiaogang and Zhou Chunya, who have gained significant recognition. Zhang's "Bloodlines–Big Family" series portrays the collective history of a generation of Chinese people, while Zhou's vibrant and colorful "Green Dog" and "Peach Blossom" series exude passionate energy and reflect the desires and aesthetics of a society undergoing dramatic changes. Their artworks have commanded astonishing prices in the art market, underscoring the flourishing Chinese contemporary art scene.

煤漏斗改造而成的时尚空间"斗廊"
"Hopper Corridor," an innovative space that repurposes the former coal-hopper unloading bridge
图片提供：大舍建筑设计事务所 *Courtesy Atelier Deshaus*

二层展厅全景　Interior view of the second floor
图片提供：大舍建筑设计事务所　*Courtesy Atelier Deshaus*
摄影：苏圣亮　*Photo by Su Shengliang*

As a private art museum built upon a personal collection, the Long Museum actively participates in the art market. However, the owners' aspirations extend beyond monetary value. They also seek to make a meaningful impact on the art world through their exhibitions. An early exhibition, "1199 People" (2014), challenged traditional curatorial philosophy and installation plans. Curator Xu Zhen curated the exhibition by selecting hundreds of pieces from the museum's collection featuring human figures and arranged them in accordance with the number of figures represented, breaking conventional notions of display and presentation.

While the overground section of the museum features the fluid exhibition space created by the as-cast-finish concrete "vault-umbrella," the galleries on the first underground floor take on a more traditional "white box" format. This particular area is dedicated to artworks from ancient times and the Republic of China period. Initially, Liu Yiqian intended to display these artworks in the overground space, but Liu Yichun, the principal of Atelier Deshaus, suggested housing them underground, explaining that treasures are often hidden under-

凝土空间，丁乙放弃了一直使用的画布材料，而特意选用了基底坚硬的木板，并首次尝试雕刻——用刀在颜料基底上刻出充满肌理的力度。为了匹配展厅的超大空间，丁乙又专门创作出近 5 米高的超大尺度的画作——这个高度早已超越了传统架上绘画的尺幅，为此他用上了电动升降机，站在上面进行创作。奥拉维尔·埃利亚松在美术馆动工之初（2013 年）前来探访，洞穴一般的内部空间让他久久驻足，2016 年他用一组全新的作品邀请观众进入"洞穴"，沉浸体验由他打造的自然乌托邦。埃利亚松用一系列几何图形，圆形、球体、立方、棱锥与美术馆的矩形展厅、弧形天顶对话，一场科学精密的美在巨大的建筑内部生产、扩散。超大展厅也为艺术家留下了"世界之最"级别的展览，2021 年的"盐田千春：颤动的灵魂"展览就成为这位艺术家最大规模的个展。

除了与当今世界做即时连接，在龙美术馆也能看到当代艺术史上教科书级别的人物。路易丝·布尔乔亚 1911 年生于法国巴黎，创作力奇崛丰沛，从 20 多岁到 90 多岁，经历了 20 世纪后半叶几乎所有的艺术思潮，超现实主义、抽象表现主义、极简主义、行为艺术、装置艺术。布尔乔亚最具代表性的作品，一只巨大的蜘蛛，出现在很多大都市的公共空间，伦敦、巴黎、纽约、东京、首尔等，成为城市地标性的公共建筑，其中最为宏大的一座被安置在毕尔巴鄂古根海姆博物馆的广场上。2018 年，在龙美术馆大厅展出的 10 米高巨型蜘蛛（作品《妈妈》），来自伦敦泰特美术馆的收藏。

除了国际艺术家，龙美术馆也呈现中国当代艺术的重要人物：比如张晓刚，他的"血缘—大家庭"系列被称为某代中国人的精神肖像；周春芽以色彩浓丽的"桃花""绿狗"为大众熟知，成为社会经济增长期民众的某种精神写照和审美映射。这两位艺术家的作品均均拍出千万高价，书写个人作品货币价值奇迹的同时也见证了中国当代艺术市场的起飞与蓬勃。作为以艺术收藏为基底的美术馆，龙美术馆自带某种货币秩序，但也做秩序之外的事情。开馆之初呈现的展览"1199 个人"（2014）打破了学院派对艺术史的书写，将该馆的藏品以"人脸"的方式形式筛选、陈列。

龙美术馆地上空间为当代艺术的展示，而地下一层是中国古代艺术珍品及民国时期艺术作品。地下展厅回归到"白盒子"式的矩形展示空间。"珍宝通常埋在地下，珍藏也更安全"，大舍主持建筑师柳亦春曾这样说服刘益谦，刘益谦起初想把这部分放在地上空间展示。

从 2012 年开馆至今（2023 年），龙美术馆两城三馆已举办了 200 多场展览，这些展览的海报贴在美术馆入口处的大厅，10 多米高的墙几乎被贴

ground for increased safety.

Within the entrance lobby of the Long Museum West Bund, a towering wall measuring over 10 meters in height is adorned with a vast collection of exhibition posters. Alongside its two other branches (Long Museum Pudong, established in 2012, and Long Museum Chongqing, opened in 2016), the museum has hosted a remarkable number of over 200 exhibitions to date. The passage of time is evident in the bolt holes visible on the walls of the overground floors, originally utilized for the casting of the as-cast-finish concrete, serving as a reminder of the construction process and adding character to the space.

The Long Museum West Bund transcends being a mere destination for art enthusiasts; it serves as a dynamic catalyst within the community. During early spring, over 160 cherry trees transform the square near Longteng Avenue into a captivating sea of blossoms. The museum is thoughtfully connected to riverside runways and skateboard parks, making it an unexpected encounter during activities such as skateboarding, jogging, or walking your dog. In the evenings, when the museum closes its doors, the coal-hopper unloading bridge and adjacent long corridors become a natural playground for skateboarders. The "Hopper Corridor" illuminates in a vibrant orange hue, resembling a lighthouse that safeguards the memories of the river.

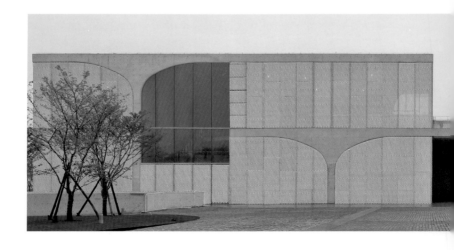

满。在地上一、二层的展厅，墙体上保留的原本用于浇筑清水混凝土的螺栓孔，也伴随着空间的使用，慢慢又留下时间的痕迹。

对于民众而言，这座美术馆不仅仅是美术馆。邻近龙腾大道广场种植的160余株樱花树每年初春都会如约绽放，成为浪漫花海。美术馆与附近的滨江跑道、滑板公园是连通的。风从黄浦江上吹来，人们或玩滑板、或牵宠物、或慢跑，浸润在这片自由温润的空气中。入夜，美术馆打烊，煤漏斗及两旁的长廊，成为滑板爱好者天然的练习场，可以从广场沿着铁轨的方向一直滑到江边。煤漏斗亮着橘色的灯，像一处平原的光盒子。

西南侧建筑立面全景　Southwestern facade
图片提供：大舍建筑设计事务所　*Courtesy Atelier Deshaus*
摄影：苏圣亮　*Photo by Su Shengliang*

设计师手绘　Sketches
图片提供：大舍建筑设计事务所　*Courtesy Atelier Deshaus*

# 诗意的巧合

口述 | 柳亦春
| 龙美术馆建筑设计师

我们设计一个美术馆，要做案例研究并研究美术馆的发展历史。最早的美术馆从哪来的？王公贵族自己有一些收藏，放在家里，请好朋友来看，慢慢诞生了欧洲的沙龙文化。最早的美术馆都是源于宫殿和别墅，比如意大利乌菲兹美术馆。画作被挂在一间间的房间内，这些房间是美术馆展厅的雏形。

就美术馆的设计而言，必须要去追究美术馆的功能跟空间的对应，它的功能究竟意味着什么？美术馆的展厅就一定是房间构成的吗？美术馆的房间相较于其他空间的房间特殊之处是什么？如此追问，就是在问美术馆的本质是什么？最终，我的思考结果是：美术馆的本质应该是一面墙。因为所有的画都要挂在墙上，所以我想这个美术馆能不能由墙体来构成，而非由房间来构成。

如果是一堵墙，那么这堵墙应该怎么布置？我意识到墙相对于房间的好处：它可以自由布局。由此人们看展的方式，就不再是从一个房间到一个房间，而是采用一个可以自由漫行的流线——这种自由感会缓解疲劳。相较于古代房间的古典平面，流动的空间也是更当代的空间。

用墙体去构筑美术馆，并且我希望这种墙体是兼顾结构性的墙体和功能性的墙体。由此诞生了"伞拱"，两把"伞"对上了之后，就形成了一个拱形的空间。搭好模型后，我倒有了点担心，因为这个拱形跟路易斯·康的金贝尔艺术博物馆有点像，建筑完成后会不会被别人说我是抄他？是否会质疑，明明是在上海，怎么做了个罗马的空间？于是我立刻就买了张机票去罗马看，卡拉卡拉浴场、哈德良离宫等，突然意识到那些空间并不只属于罗马，这样的文化已经属于全世界：这就是为什么一旦看到拱形，大家会联想到罗马。其实，

# The Poetic Coincidence

Narrated by Liu Yichun
| *Architect of Long Museum West Bund*

Before delving into the design of a museum, extensive case studies and historical research are essential. We must explore the origins of the earliest museums. In Europe, royals and aristocrats who possessed collections would invite viewers to their palaces or residences, thus giving rise to salon culture. The initial art museums often emerged from palaces or villas, such as the Uffizi Galleries in Italy. Paintings would be hung on the walls of various rooms for visitors to appreciate, effectively turning the rooms into exhibition halls.

When considering the design itself, it is crucial to seek coherence between function and space. What exactly do the functions entail? Must the exhibition halls be traditional rooms? What distinguishes museum rooms from regular rooms? These inquiries ultimately lead to one fundamental aspect: the essence of a museum. In my perspective, that essence is embodied by the wall. After all, all paintings need to be hung on walls. This realization prompted me to contemplate building the museum primarily with walls instead of rooms.

The question then arose: how should these walls be arranged? I recognized that walls, compared to rooms, offer the advantage of a flexible plan. Consequently, the museum visitation experience would no longer be a sequential contemplation of one room after another but rather a free-flowing stream, imbued with a sense of liberation that alleviates weariness.

The concept of the "vault-umbrella" structure was born from the idea of using walls, which serve both structural and functional purposes, to construct the museum. When two such umbrella-like structures intersected, an arched space was formed. However, after completing the models, I couldn't help but feel a sense of unease. The unintentional yet undeniable similarity to Louis Kahn's Kimbell Art Museum raised concerns. Would I be accused of copying his design? Furthermore, the presence of vaults and arches, reminiscent of Roman architecture, led me to ponder why I was incorporating a Roman-like space in Shanghai. To find clarity, I traveled to

中国陕北的窑洞也是拱形。追溯起来，拱形空间最早来源于山洞，人类在山洞聚居并形成洞穴文化，所以拱形是洞穴的隐喻。

2014年龙美术馆的收藏展"1199个人"，一面墙挂满了画，起初我有点惊讶，因为此前没有设想过大厅可以做这样的布展。但转念一想，美术馆的雏形，那些意大利王公贵族的私邸不正是这样吗？一面墙上布满了画。用类似于古代罗马的展览方式，在一个带着罗马属性的空间呈现，这样的巧合让我开心。对我来说，这样的巧合越多，就越会体现这是一个好的建筑。巧即"公输子之巧"，公输子是鲁班，匠人的高水平加上天作之合，这才是一种理想的建造物的状态。

每一次设计我都会追求这种巧合的状态。龙美术馆就集合了很多巧合。就如"墙伞"，它既解决了墙体结构、框架结构的连接问题，又解决了美术馆的起源、功能性的问题，还解决了空间的覆盖问题，也把管线都藏在了里面。煤漏斗是重复的，我用一把把"伞"的重复，让它们对话，完成了共同的内在逻辑建构。在历史遗存物的旁边做了一个新建筑，两者又形成了一个新的整体。

煤漏斗是龙美术馆公共开放空间的一个核心。煤漏斗是一个工业建筑、工业遗存，也是上海这座城市发展的见证，是上海历史重要的一部分。过去，当它要承担运煤的功能时，被设计成垂直于黄浦江的，这是最方便的运输路径；如今，这个垂直关系从工业与水岸的关系，演变成人们从城市抵达江边的关系，也生成了新的时空关系。

建筑模型　Model
图片提供：大舍建筑设计事务所　Courtesy Atelier Deshaus

Rome, where I explored the Baths of Caracalla, Hadrian's Villa, and other architectural marvels. It was during this journey that I had an epiphany: the association of arches with Rome stems from the fact that these vaulted spaces have transcended Rome's boundaries to become a global architectural asset. They are not exclusive to Roman architecture; for instance, similar arched spaces can be found in the cave dwellings of northern Shaanxi Province. Arches, in a metaphorical sense, originated from the caves where our ancestors sought shelter. Therefore, arches can be seen as symbolic caves.

In the 2014 exhibition "1199 People" at the Long Museum West Bund, the lobby featured an entire wall covered in paintings. I was taken aback because this display method hadn't crossed my mind during the museum's design phase. However, I pondered, why not? Isn't it fascinating to exhibit artworks in a manner reminiscent of ancient Rome, where nobles showcased their collections on walls within an architecture that bears hints of Roman influence? Such a delightful coincidence! To me, an architecture that allows for such serendipitous encounters can be considered exceptional, and the more coincidences, the better. Coincidences are like vessels of communication. When a high level of craftsmanship in design combines with poetic coincidences, it creates an ideal state for architecture.

I strive to explore these coincidences in every design endeavor. The museum embodies numerous examples, one being the "wall-umbrella." This architectural element fortuitously resolves multiple issues, including the connection between wall and frame structures, the origin and functionality of museums, spatial coverage, and a solution for concealing pipes. The coal-hopper unloading bridge, with its multiple hoppers, inspired the repetition of the "umbrella" motif, establishing a dialogue between the architecture and the bridge, as they adhere to the same construction logic. By uniting something new with something old, they form a cohesive whole.

The coal-hopper unloading bridge holds significant importance within the museum's public space. This industrial relic bears witness to Shanghai's development and is an integral part of the city's urban history. Initially designed perpendicular to the Huangpu River, it served as the most sensible route to transport coal. Now, this perpendicular line connecting the waterfront and industry has transformed into a connection from the city to the waterfront, symbolizing a new era.

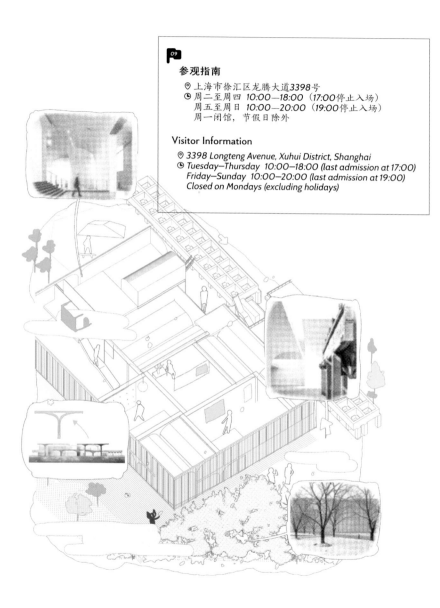

## 参观指南

◎ 上海市徐汇区龙腾大道3398号
🕙 周二至周四 10:00—18:00（17:00停止入场）
　周五至周日 10:00—20:00（19:00停止入场）
　周一闭馆，节假日除外

## Visitor Information

◎ 3398 Longteng Avenue, Xuhui District, Shanghai
🕙 Tuesday–Thursday 10:00–18:00 (last admission at 17:00)
　Friday–Sunday 10:00–20:00 (last admission at 19:00)
　Closed on Mondays (excluding holidays)

西岸美术馆沿江主入口　Main entrance to the West Bund Museum from the riverside
图片提供：戴卫奇普菲尔德建筑事务所　*Courtesy David Chipperfield Architects*
摄影　*Photo by Simon Menges*

# 10
# 西岸美术馆
# West Bund Museum

建筑师：戴卫奇普菲尔德建筑事务所 ｜ 开馆时间：2019 年
Architects: David Chipperfield Architects ｜ Opening Year: 2019

消除美术馆与公共区域的强烈边界感，把公共空间当作美术馆的附属设施，使美术馆成为更多人可以走得进去的场所，是西岸美术馆对于美术馆空间的理解。

The West Bund Museum goes beyond being a mere art destination by incorporating public facilities that enhance the overall experience. This approach emphasizes the museum's dedication to inclusivity and encourages anyone to engage with art and culture.

沿龙腾大道，自北向南，中间横隔一片水域——这是龙华港与黄浦江交汇的地方。若从西岸美术馆大道的北端出发，水域上方的龙华港桥是去往西岸美术馆的必经之路。桥上有奔忙的车辆，也有闲适的步行者，桥身几何形的镂空设计既能遮风挡雨，也让人与自然相连。常有奔劳的骑行者，在桥上停下，望一望江水、滩涂、低飞的水鸟。桥的南侧是徐汇滨江的海事塔广场，海事塔是这片滨江景观的最高点。

如果说龙华港桥以北的几座美术馆如珍珠般连缀成了一条线状的艺术带，那西岸美术馆则是西岸美术馆大道南段景观的核心——它的西侧是西岸艺术中心，每年西岸艺术与设计博览会的举办场地；西北方是由近 20 处建筑构成的艺术社区，入驻了多家知名画廊、建筑机构、艺术家工作室；沿江往南是油罐艺术中心。

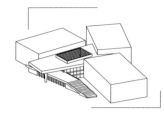

Starting from the northern end of the West Bund Museum Mile along Longteng Avenue, you need to walk southwards, passing the Longhua Gang Bridge, in order to reach the West Bund Museum. The bridge stretches over a waterway where Longhua Gang converges with the Huangpu River. With its stunning geometric trusses, the double-deck bridge efficiently divides vehicular and pedestrian traffic. Standing at the southern end of the bridge is the Maritime Watchtower, a captivating landmark that was once a beacon but has now assumed a remarkable role by the riverside.

Take the Longhua Gang Bridge as the reference point, the northern portion of the West Bund Museum Mile features a collection of museums interconnected like a string of pearls. However, the southern portion of the mile is centered around the West Bund Museum. In its vicinity, you can find the West Bund Art Center towards the west, which serves as the venue for the annual West Bund Art & Design Fair. To the northwest, there is an art community comprising 18 buildings, housing an array of galleries, architectural firms, and artist studios. Moving further south, you will come across TANK Shanghai.

The exceptional role played by the West Bund Museum is primarily attributed to its remarkable openness. The museum, designed by renowned British architect David Chipperfield, who was awarded the prestigious 2023 Pritzker Prize, is situated on a triangular plot located at the northernmost point of a newly established public park. Positioned at the convergence of Longteng Avenue and the Huangpu River, the building is elevated above the flood plain, and a raised public esplanade surrounds it, granting panoramic views of the river.

On the eastern side, the edge of the esplanade is defined by a continuous sequence of steps that lead to the riverbank, incorporating landing stages. This design feature enhances accessibility and provides improved connections between the museum, the river, and the park. The intention behind this design is to transform the museum into an inclusive public space, accessible and welcoming to all.

The West Bund Museum is composed of three primary gallery volumes arranged in a pinwheel configuration, encircling a central lobby that features a double-height atrium. By purposefully twisting the positioning of these three volumes, striking interstitial public spaces are created within the building.

The central lobby, serving as a focal point, can be reached from two different directions. Access from Longteng Avenue is available from the west, while access from the riverside is provided from the east. The riverside entrance offers two options: visitors can either descend into a sunken courtyard located at a lower level or ascend a set of stairs from the esplanade to reach the central lobby at a higher level. Both of these levels provide entry to the primary gallery spaces within the museum.

北侧视角 & 沿街入口　View from the north & Entrance from the street
图片提供：戴卫奇普菲尔德建筑事务所　Courtesy David Chipperfield Architects
摄影 Photo by Simon Menges

The West Bund Museum goes beyond being a mere art destination by incorporating public facilities that enhance the overall experience. After exploring the exhibitions, many visitors are drawn to the steps on the east side of the esplanade, leading to the riverbank, where they can reflect and appreciate the scenic river view. The ground floor of the museum is thoughtfully designed as a public thoroughfare, seamlessly connecting the urban street life with the captivating vistas of the Huangpu River. At the east entrance, a riverside café welcomes both visitors and the general public, offering longer opening hours than the museum itself. This integration of public spaces encourages a sense of community and provides a space for visitors to relax, enjoy refreshments, and immerse themselves in the vibrant atmosphere of the riverside.

The West Bund Museum embodies a rational and humble demeanor, skillfully incorporating natural light and rational lines within its space. The interplay of daylight streaming through the structure creates captivating light and shade effects, accentuating the gallery spaces. When needed, certain areas can be illuminated by natural light. Despite its understated exterior, the initial plan to display the museum's logo on the facade was abandoned to preserve the overall harmony and balance of the building. This humility is intentional, aiming to shift the focus away from the design itself and towards contemplation and the

韩国艺术家梁慧圭的作品《萦绕的智性》在一层中庭展示,2021—2023
"Lingering Nous" by Korean artist Haegue Yang was exhibited in the atrium, 2021-2023
图片提供:西岸美术馆 *Courtesy West Bund Museum*

成为景观核心的前提是建筑的开放性。作为戴卫·奇普菲尔德及其团队的优秀作品，该建筑不仅是矗立于项目基地上的一座独立建筑，还扮演着优化基地与黄浦江、公园之间联系的公共角色。除了一个开阔的向河岸开放的广场，设计师还极富创意地将三个主要的展厅呈风车状布置——这样的布局方式让美术馆内部的中庭空间可以更多地向城市空间打开，建立了大众与空间的连通。这种开放性还体现在进入美术馆的路径。不同于传统博物馆的单一路径，这里提供了三种路径，灵活、便捷：从西侧龙腾大道和东侧滨江步道进入；或从滨江一侧下沉庭院进入；或从滨江一侧的广场台阶行至二层露台再进入中央门厅。

开放性的设计，让该建筑不仅完成了作为美术馆的功能和使命，还让其兼具城市客厅、休息驿站的功能。很多观众看完展后会顺着广场东边的台阶走到江边，坐着吹吹风——这里有专门为人休憩设置的椅子。美术馆临江那侧的大门里，咖啡厅向公众开放，营业时间比美术馆开放时间更长，即使美术馆闭馆，也可以进来喝一杯，吃一份画作限定版甜品。

消除美术馆与公共区域的强烈边界感，把公共空间当作美术馆的附属设施，使美术馆成为更多人可以走得进去的场所，是西岸美术馆对于美术馆空间的理解。而这种开放性也体现在展厅：西岸美术馆一层中庭和 0 号展厅是免费向公众开放的。一层中庭是公共的阅读空间，也是一家书店。书店是专业做艺术类图书的品牌 BOOK ART，书店品牌在视觉上做了弱化，没有常规店铺的门头、招牌——这让其完美融合在美术馆的公共空间里，成为美术馆的一部分。各类精美的画册、原版书籍在书架上，供人翻阅；有椅子，可以坐下慢慢看。中庭空间也会成为艺术品的展示空间，与书架、座椅一道，浑然一体，成为空间的一部分——此刻，艺术品消解了高高在上、被人观看的地位，成为日常。韩国艺术家梁慧圭的作品《萦绕的智性》自上而下悬挂，暗粉色和绿色的百叶窗、LED 灯——都是寻常可见的家装材料——乍一看还以为是中庭的装饰。《萦绕的智性》是展览"万物的声音"中的一件作品，该展作为常设展，展期长达近两年。近两年以来，这件作品一直与来来往往的过客相伴。[1]

戴卫·奇普菲尔德是 2023 年普利兹克奖得主，这位优雅的英国爵士，作品也散发着理性、谦虚的绅士气质。"空间线条是理性的，横平竖直；巧妙引用了自然天光，天晴时，你会看到随着建筑结构打洒下来的光在中庭形

---

[1] 该作品已于 2023 年 5 月撤展。

human experience. From the moment visitors step into the museum's lobby, a sense of comfort and serenity envelops them, setting the tone for immersive exploration and reflection.

In line with the West Bund Museum's commitment to creating an accessible public space, both the central lobby and Gallery 0 are open to the public free of charge. This approach emphasizes the museum's dedication to inclusivity and encourages anyone to engage with art and culture. Within the ground-floor atrium, visitors will find a bookstore operated by BOOK ART, a renowned brand specializing in art books. The absence of attention-seeking signs or logos is deliberate, allowing the bookstore to quietly welcome visitors. The shelves are adorned with exquisite art catalogs and original books, providing a serene and enriching experience for book enthusiasts and art lovers alike.

The atrium serves as a distinctive exhibition space, showcasing intriguing and captivating installations. For nearly two years, visitors had the opportunity to view a large-scale blind installation titled "Lingering Nous" in iridescent green and pink, suspended from above. This striking artwork was created by the renowned Korean artist Haegue Yang, known for her innovative use of everyday objects and industrially produced materials. Originally commissioned for the Centre Pompidou in Paris, "Lingering Nous" was featured in the West Bund Museum as part of the semi-permanent exhibition titled "The Voice of Things: Highlights of the Centre Pompidou Collection Vol. Ⅱ " (July 28, 2021–May 7, 2023). This exhibition marked the second chapter of the collaborative project between the Centre Pompidou and the West Bund Museum, known as the "Centre Pompidou × West Bund Museum Project."

The project was officially launched in November 2019, coinciding with the inauguration of the West Bund Museum, and was honored by the presence of French President Emmanuel Macron. This collaborative endeavor emerged from a Memorandum of Understanding signed in 2017 between the Centre Pompidou and the Shanghai West Bund Development Group, establishing a cultural partnership between the institutions. Widely regarded as a significant initiative for cultural exchange and cooperation between China and France, the project was even included in the Joint Declaration between the People's Republic of China and the French Republic in 2018.

Over the course of five years following the museum's opening, the project entails a series of jointly curated semi-permanent exhibitions and feature exhibitions on display at the West Bund Museum. Additionally, it encompasses a range of activities such as seminars, performing arts events, educational initiatives, and cultural outreach. Furthermore, the Centre Pompidou in Paris will host projects and exhibitions showcasing the work of Chinese artists, fostering cross-cultural artistic dialogue and appreciation between the two countries.

The Centre Pompidou × West Bund Museum Project commenced with its

二层展厅　Upper gallery
图片提供：戴卫奇普菲尔德建筑事务所　*Courtesy David Chipperfield Architects*
摄影　*Photo by Simon Menges*

西岸美术馆与蓬皮杜中心五年展陈合作项目

1　"万物的声音：蓬皮杜中心典藏展（二）"展览现场，2021
2　"抽象艺术先驱：康定斯基"展览现场，2021
3　"胡晓媛：沙径"展览现场，2021
4　"本源之画：超现实主义与东方"展览现场，2023
5　"姚清妹：鼹鼠"展览现场，2023
6　"她们与抽象"展览现场，2022—2023

图片提供：西岸美术馆 *Courtesy West Bund Museum*

Centre Pompidou × West Bund Museum Project

1　The Voice of Things: Highlights of the Centre Pompidou Collection Vol. II , 2021
2　Kandinsky: The Pioneer of Abstract Art, 2021
3　Hu Xiaoyuan: Paths in the Sand, 2021
4　Painting the Essential: Surrealism and the East, 2023
5　Yao Qingmei: Mole, 2023
6　Women in Abstraction, 2022–2023

成很好看的光影；根据展陈需要，展厅里面也有一些空间，可以进入自然光。"美术馆的工作人员站在一层大厅，介绍道："大卫作品的空间在外观看比较低调，起初我们与事务所沟通提出，是否要在建筑上呈现美术馆的 logo，但设计方觉得 logo 会破坏建筑本身的平衡性、整体感，所以现在的空间并没有用很大的 logo。"谦虚的设计是让人舒适的，它的存在不是为了彰显设计本身，而是处处以人去思量；这种舒适感，自踏入美术馆大厅那刻就能感受到，身体的舒适、观看的舒适、行走的舒适……空间与光的配比，刚刚好；空间与身体的尺度，刚刚好；气息的流动，刚刚好。

2019 年 11 月，西岸美术馆开馆，法国总统埃马纽埃尔·马克龙亲临现场，为"西岸美术馆与蓬皮杜中心五年展陈合作项目"（下文简称五年合作项目）揭幕。该项目作为"中法最高级别文化合作项目"于 2018 年被纳入《中法联合声明》。合作包括：在未来五年内，双方将以共同策划为前提，在西岸美术馆展开常设展、特展；同时在法国巴黎蓬皮杜中心也将呈现聚焦中国现当代艺术景象的展览等文化项目；双方也将共同组织演出、研讨、公共教育等活动。

开馆展即为"五年合作项目"的开始，展期长达一年半的常设展"时间的形态：蓬皮杜中心典藏展"在万众瞩目中登场。2021 年 5 月，在"抽象绘画的先驱"康定斯基诞辰 155 周年之际，特展"抽象艺术先驱：康定斯基"开幕。两个月后，黄浦江对岸的浦东美术馆开幕。像是一场奇妙的文化共振，浦东美术馆设计建造的灵感来源于抽象艺术的另一位大师马列维奇。"点、线、面"的艺术，如此清晰、具体、全面、生动地呈现在中国观众面前。

"五年合作项目"不仅带来了 20 世纪艺术史的经典作品，成为难得的教科书般的展陈，也有中国艺术家的系列呈现。位于一层的 0 号展厅，自 2021 年持续开展中国年轻艺术家项目，于吉、陈维、胡晓媛、姚清妹等，这些正值事业上升期的艺术家用扎实的创作，为人们呈现了一个个具象的"中国当代艺术"。值得一提的是，作为免费向公众开放的 0 号展厅，这些展览不仅做得精细，连免费的随展画册都很精美。艺术家胡晓媛在此的作品《沙径》，展览画册与作品材质有了巧妙呼应。绡（生丝）是一种年代久远的传统材料，有着蝉翼般透明的质感，这是胡晓媛多年来一直研究和使用的材料。随展画册中使用了和绡有着相似质感的宣纸，轻薄、透光、泛黄的宣纸上印着与展览相关的诗歌，以插页的形式，夹在画册中；与画册的尺寸不相同，薄薄的宣纸叠着，像一封书信。这样的巧思让人感动，是观展的意外之喜。

first semi-permanent exhibition, "The Shape of Time: Highlights of the Centre Pompidou Collection vol. Ⅰ " (November 8, 2019–May 9, 2021), serving as the inaugural showcase. It was followed by the second chapter, titled "The Voice of Things." In a coincidental alignment, the project presented "Kandinsky: The Pioneer of Abstract Art" (May 1–September 19, 2021) on the 155th anniversary of Kandinsky's birth. This exhibition marked the third feature exhibition within the series. Intriguingly, two months later, the Pudong Art Museum, an architectural marvel inspired by abstract artist Kazimir Malevich, opened its doors on the opposite side of the Huangpu River. The simultaneous presence of the West Bund Museum and the Pudong Art Museum created an abstract art duet, providing a captivating and enriching experience for the local audience.

This project not only serves as a platform for showcasing the renowned Centre Pompidou collection, offering insights into the history of art in the 20th and 21st centuries, but it also places a significant emphasis on promoting Chinese contemporary art through a series of projects. Starting from 2021, Gallery 0 has been exclusively dedicated to highlighting the works of young Chinese artists.

Notable artists such as Yu Ji, Chen Wei, Nabuqi, Hu Xiaoyuan, and Yao Qingmei have had the opportunity to present their unique interpretations of Chinese contemporary art within this space. Among these exhibitions, Hu Xiaoyuan's "Paths in the Sand" (November 11, 2022–March 12, 2023) stands out for its clever integration of the exhibition catalog with the artwork itself, which utilizes raw silk as a primary material. The pliability and translucency of the raw silk create a subtle and tactile experience for viewers. The exhibition catalog also incorporates a type of Xuan paper that shares a similar texture with raw silk—thin, lightweight, and slightly yellowish. This delicate paper is folded meticulously, resembling a long-awaited letter, and features printed poems that are closely related to the artwork. This thoughtful synergy between the catalog and the artwork enhances the overall immersive and contemplative experience for the reader.

The Centre Pompidou × West Bund Museum Project serves as a catalyst for enhanced cooperation and cultural exchange between China and France, presenting a captivating tapestry of shared artistic endeavors. Showcasing vibrant details of cultural exchange between the two countries, the exhibition "Painting the Essential – Surrealism and the East" (April 29–September 24, 2023) offers a fresh perspective on Surrealism, focusing on its abstract tendencies and previously unexplored dialogue with the East. The exhibition unveils intriguing connections, such as André Masson's utilization of the dynamic and fluid characteristics of Chinese calligraphy as inspiration for his groundbreaking series of "automatic paintings." Additionally, the exhibition explores Henri Michaux's creation of a series of ink paintings in 1958 and the experiences of Chinese artists like Zao Wou-Ki, who embarked on journeys to Paris, fostering friendships and artistic

"五年合作项目"作为中法文化的交流，并未只停留在将艺术品在巴黎与上海两地之间挪移，而是呈现了中法文化交流的生动细节。展览"本源之画：超现实主义与东方"，呈现了20世纪上半叶欧洲艺术家与中国及东方艺术的彼此影响，比如安德烈·马松借鉴了中国书法灵动迅捷的特点，创作了第一批"自动绘画"；亨利·米肖于1958年创作了数幅水墨画；同时赵无极等一批中国艺术家到巴黎去学习和创作，展开了与欧洲艺术家们的友谊。

　　值得一提的是，"五年合作项目"还将舞台特意给予了女性艺术家。在以男性书写主导的当代艺术史中，女性艺术家一直是被忽略、被匿名的。特展"她们与抽象"呈现了女性对抽象艺术的重要贡献，涉及的艺术作品横跨三个世纪，涵盖了抽象表现主义、光学与动能艺术、纤维艺术等各种流派形式，展览亦涵盖了中国的女性艺术家。该展并非孤例，西岸美术馆一直有着对女性创作力的关注和支持：在0号展厅持续展陈的中国艺术家项目，自2021年至2023年的五位艺术家中有四位是女性艺术家；"跃动她影在西岸"为女性舞者提供了持续的舞台和支持……这些都共同谱写了关于女性主义、女性叙事的动人篇章。

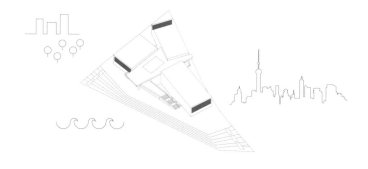

视野图解　　Volume diagram, views
图片提供：戴卫奇普菲尔德建筑事务所·Courtesy David Chipperfield Architects

exchanges with European counterparts. Drawing from the esteemed Centre Pompidou collection and enriched by loans from the Shanghai Museum, this exhibition stands as a testament to the remarkable cultural heritage and ongoing dialogue between China and France.

The feature exhibition "Women in Abstraction" (November 11, 2022–March 8, 2023) offered a compelling reassessment of the significant contributions made by numerous female artists. As part of the Centre Pompidou × West Bund Museum Project, curator Christine Macel presented a unique reinterpretation of the exhibition, drawing from the rich holdings of the Centre Pompidou collection. The showcased artworks spanned three centuries, encompassing diverse forms and genres, such as abstract expressionism, optical and kinetic art, and fiber art. Importantly, the exhibition also featured the works of Chinese female artists, further broadening the narrative of women's role in abstraction. Through this curated showcase, "Women in Abstraction" shed light on the remarkable creativity and artistic achievements of women throughout history, fostering a greater appreciation for their contributions within the art world.

Dedicated to promoting gender equality and supporting female artists, the West Bund Museum has consistently championed the work of women in the arts. Notably, four out of the five Chinese artists featured in Gallery 0 between 2021 and 2023 were female, showcasing the museum's commitment to providing a platform for their creative voices. In collaboration with Kering, the Centre Pompidou × West Bund Museum Project launched "Women In Motion at West Bund" in 2021, a program designed to celebrate women's creativity in dance and choreography. This initiative serves as a powerful platform for challenging tradi-

tional mindsets and fostering reflection on the place and recognition of women across artistic disciplines. The second edition of "Women In Motion at West Bund" took place in April 2023, further amplifying the artistic contributions and accomplishments of women in the field. Through these efforts, the West Bund Museum and the Centre Pompidou × West Bund Museum Project actively contribute to shaping a more inclusive and equitable landscape for female artists and their invaluable contributions to the arts.

**参观指南**

⊚ 上海市徐汇区龙腾大道2600号
⊙ 周二至周日 10:00—17:00（最后入场时间 16:30）
  周一闭馆（包括展厅、书店及公共空间）

**Visitor Information**

⊚ 2600 Longteng Avenue, Xuhui District, Shanghai
⊙ Tuesday–Sunday 10:00–17:00
  (last admission at 16:30)
  Closed on Mondays, including all public facilities

二层露台江景　View of the Huangpu River from the second-floor terrace
图片提供：戴卫奇普菲尔德建筑事务所　*Courtesy David Chipperfield Architects*
摄影　*Photo by Simon Menges*

油罐艺术中心全景　Panoramic view of TANK Shanghai
图片提供：油罐艺术中心　*Courtesy TANK Shanghai*

# 11
# 油罐艺术中心
# TANK Shanghai

建筑师(改造):OPEN 建筑事务所 | 开馆时间:2019 年
Architects: OPEN Architecture (renovation) | Opening Year: 2019

"始于艺术,不囿于艺术",
让美术馆成为如乐园一般的存在。

By organizing a rich array of interdisciplinary performances and events, TANK Shanghai embraces the concept of originating from art while not being limited to it.

　　从西岸美术馆再往南,五个白色的圆柱形储油罐成为西岸美术馆大道的完美句号。这五个储油罐原隶属于中国最早建成的机场——上海龙华机场(1966 年停用)。油罐内部空间开阔,挑高 15 米以上,拥有极具工业特点的穹顶及曲面圆周空间。经由 OPEN 建筑事务所六年的改造,五个储油罐重获新生,成为一个综合性的艺术中心。五个油罐中,三个油罐连通,成为美术馆的主体空间,它们各自独立又由美术馆大厅相连。其中,3 号罐的穹顶上开了天窗,光线能从上方透入展厅;4 号被分割成三层;5 号罐面积最大,内部被分成上下两层,每层面积超过 600 平方米。除五个油罐外,艺术中心还设置了两个开阔的广场。

　　上海油罐艺术中心由藏家乔志兵创办,该馆展览体现了这位世界级藏家的视野和趣味。比如,阿德里安·维拉·罗哈斯,这位以全球游牧为方式工

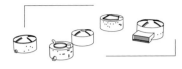

Located to the south of the West Bund Museum, the West Bund Museum Mile concludes with a striking landmark: five white cylindrical oil tanks. Originally part of the Shanghai Longhua Airport, one of China's earliest airports that ceased operations in 1966, these tanks have undergone a comprehensive six-year renovation by the Beijing-based OPEN Architecture to become TANK Shanghai. This dynamic contemporary art center now houses galleries, public spaces, green areas, plazas, cafes, restaurants, and more. The adaptive reuse of the tanks has breathed new life into these historically significant structures, creating a vibrant cultural destination.

Each of the five tanks at TANK Shanghai has a unique program and exhibition strategy. Tanks 3, 4, and 5 are interconnected and serve as the primary exhibition space, while the remaining two tanks fulfill supporting functions.

Tank 3 features an open skylight that allows direct light and even rain to enter the space. Tank 4 has a square layout across three floors, offering galleries specifically designed for paintings and sculptures. Tank 5, the largest tank, includes two projecting rectangular volumes that serve as expansive stages for open-air performances. Each stage covers an area of over 600 square meters, providing ample space for captivating and immersive performances.

Founded by collector Qiao Zhibing, TANK Shanghai reflects the owner's personal interests and visionary approach through its exhibitions. The art center debuted in 2019 with Argentinean artist Adrián Villar Rojas' solo exhibition, "Sometimes you wonder, in an interconnected universe, who is dreaming who?". This exhibition set the tone for Tank Shanghai's engaging and thought-provoking programming.

The curved interior spaces of the tanks are ideal for showcasing visually and acoustically immersive multimedia works. For example, teamLab's "Universe of Water Particles in the Tank" (2019) transformed Tank 5 into an interactive and ever-changing installation, mesmerizing visitors with its captivating effects.

In the 2021 exhibition "Theaster Gates: Bad Neon," the American artist recreated a 1980s disco atmosphere by utilizing the unique structure of TANK Shanghai as a roller-skating rink. Visitors were invited to engage with contemporary art through movement, sound, and a new series of neon works inspired by abstract painter Agnes Martin and social theorist W.E.B Du Bois. This immersive experience combined different artistic elements to create a captivating and memorable encounter.

TANK Shanghai fosters a playful and interactive atmosphere through various activities that invite public participation. The annual TANK Art Festival, launched in 2019, combines art exhibitions, book fairs, markets, food and beverages, music, urban culture, and performances. Esteemed experts from

作的阿根廷青年艺术家，在这里举办了在中国的首次美术馆个展。曲面圆周的内部空间也特别适合呈现有着强烈冲击的视听类的多媒体作品，让观众充分沉浸式体验。2010年后火爆全球的艺术团体teamLab在5号罐呈现了"油罐中的水粒子世界"。油罐的特殊空间给予艺术家灵感，西斯特·盖茨把这里打造成有着20世纪80年代迪厅效果的旱冰场，邀请观众以轮滑的方式感受艺术。

除了持续呈现国内外艺术家的重要展览，上海油罐艺术中心还用一系列的活动增加"玩"的氛围感，邀请大众一起"玩"。比如2019年创办的"玩家艺术节"，让艺术家走出严肃的学院氛围，成为如同市集摊主的存在，用作品装饰自己的摊位，与买家热络聊天。每年一次的"玩家艺术节"也成为全城时髦青年人的派对。此外，这里还持续推出了各种跨界演出和活动，"始于艺术，不囿于艺术"，让美术馆成为如乐园一般的存在。

入口广场成了演出和活动的理想场所
Entrance square becomes the perfect venue for events and performance
图片提供：油罐艺术中心 *Courtesy TANK Shanghai*

上图:"凸面/凹面:比利时当代艺术展"展览现场,2019
下图 & 右图:夜景
Above: Convex/ Concave: Belgian Contemporary Art, 2019
Below & Right: Night view

图片提供:油罐艺术中心 *Courtesy TANK Shanghai*

diverse fields create a dynamic cultural experience for the public. Artists become stallholders, decorating their booths with artworks and engaging in lively conversations with visitors. Additionally, TANK Shanghai hosts the lively TANK New Year's Eve party, celebrating the arrival of the new year in a vibrant artistic setting.

By organizing a rich array of interdisciplinary performances and events, TANK Shanghai embraces the concept of originating from art while not being limited to it. The art center transforms into a welcoming playground, inviting everyone to participate and engage in cultural exploration and enjoyment.

**参观指南**

- 上海市徐汇区龙腾大道2380号
- 周二至周五 *12:00—18:00*（17:30停止入场）
  周六、周日 *10:00—18:00*（17:30停止入场）
  周一闭馆

## Visitor Information

- 2380 Longteng Avenue, Xuhui District, Shanghai
- Tuesday—Friday *12:00—18:00* (last admission at 17:30)
  Saturday—Sunday *10:00—18:00* (last admission at 17:30)
  Closed on Mondays

热闹的广场　　The bustling square
图片提供：油罐艺术中心　*Courtesy TANK Shanghai*

2000年，M50园区从老厂房转型为艺术园区，成为苏州河沿线的重要艺术地标。此后，OCAT（上海馆）、UCCA Edge、上海苏富比空间、没顶画廊等艺术空间不断完成着苏州河艺术生态的勾勒。

上海最为知名的艺术聚落 M50 创意园位于苏州河上游；专注于学术生产的 OCAT 上海馆，自 2012 年成立后一直在苏州河沿线活动。UCCA Edge 的落地完成了苏州河艺术地带的勾连，成为了新的故事。

In the year 2000, M50 underwent a transformation from old factories into a creative park, becoming a significant artistic landmark along the Suzhou Creek. Subsequently, art spaces such as OCAT Shanghai, UCCA Edge, Sotheby's Shanghai, and Madeln Gallery have continuously shaped the artistic ecosystem along the Suzhou Creek.

The M50 Creative Park is situated upstream along the Creek; OCAT Shanghai, with a focus on academic production, has been active along the Creek since its establishment in 2012. The establishment of UCCA Edge completes the artistic zone along the Suzhou Creek, marking a new chapter in the area's artistic narrative.

上海西郊是成熟的大型生活区，有动物园、大型居民住宅区、超大商业综合体。明珠美术馆坐落在爱琴海购物公园内，是商场里的美术馆。

The western suburbs of Shanghai boast a thriving and well-established residential area that includes a zoo, spacious residential neighborhoods, and a sprawling commercial complex. Nestled within the Aegean Place, the Pearl Art Museum serves as an art gallery within this vibrant shopping mall.

# 其他区域
## Other Areas

1号线地铁上海马戏城站以北,是过去的老工业厂区,这里有兴建于20世纪五六十年代的一批厂房及大型工人新村。明当代美术馆就坐落于此,曾被誉为"花园工厂"的上海造纸机械厂的车间,如今成为当代艺术的生产地。

North of Shanghai Circus Station on Line 1 of the subway lies the former old industrial zone, which housed a collection of factories and large-scale workers' villages constructed in the 1950s and 1960s. Among them, the Ming Contemporary Art Museum stands prominently, taking residence in the workshop of the Shanghai Paper Machinery Factory, fondly referred to as the "Garden Factory." Today, this transformed space serves as a thriving hub for the production of contemporary art.

崇明岛,中国第三大岛屿,由长江夹带下来的泥沙冲击而成。在崇明岛的最西边,农田深处,没顶美术馆正在进行岛上的"艺术开垦"。

Chongming Island, the third largest island in China, was formed by the sediment deposited by the Yangtze River. At the westernmost tip of Chongming Island, nestled deep within the farmlands, the MadeIn Art Museum is dedicated to the "art cultivation" of the island.

盈凯文创广场外景，UCCA Edge位于二层至四层的裙房内
Exterior of EDGE. UCCA Edge is located on the second to fourth floors of the building
摄影：倪铭杰 *Photo by Ni Mingjie*

# 12
# UCCA Edge

内部空间设计：SO-IL 建筑事务所 ｜ 开馆时间：2021 年
Interior Design: SO-IL ｜ Opening Year: 2021

UCCA Edge的落地完成了苏州河艺术
地带的勾连，成为了新的故事。

The arrival of UCCA Edge completes
the art belt along the Suzhou Creek,
further enriching Shanghai's vibrant art scene.

　　位于北京 798 艺术区的 UCCA 尤伦斯当代艺术中心（简称 UCCA）建于 2007 年，是中国最早一批建立并活跃至今的大型艺术机构，由世界重要的艺术品藏家尤伦斯夫妇创办。UCCA 不仅是北京知名艺术地标，也是中国当代艺术进程的重要参与者、发生场。2017 年 UCCA 完成机构重组，转型为 UCCA 集团，由此该机构以美术馆群落的方式在全国生长：2018 年底，在紧邻北京的北戴河大型休闲度假社区落地 UCCA 沙丘美术馆；三年后，上海空间在苏州河畔落地。

　　与北京 UCCA 的工业遗迹空间不同，上海空间栖身于簇新的写字楼，周边地块也均是商务属性，而非北京 798 这样的艺术区。虽身处商业体的空间，UCCA Edge 还是在艺术叙事脉络上找到了与上海联通的脉络：苏州河。苏州河不仅是一个地理名词，也是具有丰富文化意味的地标。娄烨的

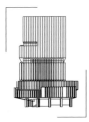

UCCA Center for Contemporary Art, founded in 2007 by Guy and Myriam Ullens as the Ullens Center for Contemporary Art in Beijing, now stands as China's foremost institution for contemporary art. In 2017, it evolved into the UCCA Group under new patrons and shareholders. UCCA offers diverse exhibitions, public programs, and research initiatives across its three locations. UCCA Beijing holds a central position in the 798 Art District, becoming a prominent landmark in the city's art scene. UCCA Dune, designed by Open Architecture and opened in late 2018, is nestled beneath the sands of the seaside enclave of Aranya in Beidaihe. UCCA Edge, designed by New York-based architecture firm SO–IL, opened its doors in Shanghai in May 2021.

In contrast to UCCA Beijing's transformed industrial heritage space in Beijing's 798 Art District, UCCA Edge resides in a modern office building surrounded by commercial properties in Shanghai. Despite this setting, UCCA Edge forges a meaningful connection with Shanghai's art narrative through the Suzhou Creek, holding geographical significance and rich cultural meaning. The renown of Lou Ye's film *Suzhou River* (2000), a tragic love story set in contemporary Shanghai, has made the Creek a popular destination for art enthusiasts. Additionally, the Creek is close to M50 Creative Park, one of Shanghai's prominent art zones, and OCAT Shanghai, the first branch of OCT Contemporary Art Terminals (inaugurated in 2012, temporarily closed in 2023). The arrival of UCCA Edge completes the art belt along the Creek, further enriching Shanghai's vibrant art scene.

The primary mission of UCCA Edge is to foster a localized art narrative. Its inaugural exhibition, "City on the Edge: Art and Shanghai at the Turn of the Millennium" (May–July 2021), delved into the pivotal moment when China's art world embraced its place within the global contemporary art scene. Around the year 2000, amidst emerging markets, institutional reforms, and artist-led initiatives, a series of exhibitions pushed the boundaries of experimental art in a city on the verge of gaining new international significance. Notably, the renovation of old factory buildings in M50 Creative Park around the year 2000 played a significant role in transforming it into an art zone.

UCCA Edge artfully occupies the second to fourth floors of the building, ingeniously transformed into an art museum. The exhibition area features museum-grade temperature and humidity control, ensuring the preservation of artworks. Carefully designed, it can accommodate diverse exhibition requirements, with double-height spaces for showcasing large-scale installations or sculptures.

The fourth-floor wraparound outdoor terrace also serves as a platform for displaying installations and sculptures. For the exhibition "Urban Theater: A Comedy in Four Acts" (May 2021–May 2022), three site-specific artworks were created by Cao Fei, Aki Sasamoto, and Wong Ping, taking inspiration from the

欧文·沃姆的户外装置作品《UFO》在UCCA Edge的底层入口处展出，2021
Erwin Wurm's artwork "UFO" was exhibited at the ground floor entrance of UCCA Edge in 2021
图片提供：UCCA尤伦斯当代艺术中心 *Courtesy UCCA Center for Contemporary Art*

"成为安迪·沃霍尔"展览现场,2021　Becoming Andy Warhol, 2021
图片提供:UCCA尤伦斯当代艺术中心　Courtesy UCCA Center for Contemporary Art

黄炳的户外装置作品《闪缩练习》在UCCA Edge的南侧露台展示出,2021
Wong Ping's artwork "Shifty Eyes Exercise" was exhibited on the south terrace of UCCA Edge in 2021
图片提供:UCCA尤伦斯当代艺术中心　Courtesy UCCA Center for Contemporary Art

电影《苏州河》2000年上映后，让全世界的文艺青年把苏州河当成旅行打卡点；上海最为知名的艺术聚落M50创意园位于苏州河上游；专注于学术生产的OCAT的上海馆，2012年成立后一直在苏州河沿线活动（注：该机构于2023年暂时结束）。UCCA Edge的落地完成了苏州河艺术地带的勾连，成为了新的故事。

　　上海空间落地后的首要使命是开启在地化的艺术叙事，而非成为北京空间的地理平移。开馆展"激浪之城：世纪之交的艺术与上海"，将叙事精准地落在2000年。2000年前后，随着新兴市场的发展、机构改革的深入和艺术家自发性组织的活跃，上海涌现出一系列推动艺术传播与提升艺术影响力的展览，这些展览体现并扩展了实验艺术呈现的可能性。值得一提的是，作为苏州河艺术叙事重要一环的M50园区，即是在2000年从老厂房转型为艺术园区。

　　UCCA Edge利用了建筑的二至四层，商用空间经由设计改造后，拥有了美术馆的功能和属性：空间规划了不同功能的展厅，比如双层高的空间用以陈列大型的装置或雕塑作品；展览全区域实现美术馆级恒温恒湿；就连四层360度的环绕型户外露台，也可呈现装置和雕塑——2021年，艺术家曹斐、黄炳、笹本晃就在这个露台展开艺术实践，用或诙谐或冷静的作品与上海展开对话。

　　UCCA美术馆群落实现了资源联动，一些展览会在北京、上海两地接力。UCCA Edge开馆后的首位艺术家个展"刘小东：你的朋友"，对艺术家自2010年北京UCCA"金城小子"个展之后10余年间的艺术实践和创作发展脉络进行集中呈现与梳理。集结了毕加索、马蒂斯、贾科梅蒂、塞尚等6位20世纪现代艺术巨匠的展览"现代主义漫步：柏林国立博古睿美术馆馆藏展"，于2023年6月抵临上海，年底将巡展至北京。相较于北京空间，上海空间的优势是交通便捷，出地铁即可直达展厅。这让逛美术馆成为一件"顺手"的事情，白领们在午休间隙即可来此欣赏艺术。

terrace's architectural attributes. Together with an outdoor installation by Erwin Wurm, they presented four comedic acts to the public in an elevated "urban theater," seamlessly woven into the cityscape with the city as its backdrop. Embracing the spectacle of the megacity, the four artists explored the humor and philosophy that underlie daily life, presenting a range of comedic acts from cheeky to absurd.

UCCA has successfully coordinated resources, allowing certain exhibitions to be showcased in both Beijing and Shanghai. UCCA Edge's debut solo exhibition, "Liu Xiaodong: Your Friends" (August–October 2021), traced Liu's artistic journey over the past decade since his 2010 exhibition "Hometown Boy" at UCCA Beijing, presenting over 120 new and existing works. Furthermore, the exhibition "Modern Time: Masterpieces from the Collection of Museum Berggruen / Nationalgalerie Berlin" will travel to UCCA Beijing in November 2023, following its display at UCCA Edge from June to October 2023. This captivating exhibition explores the evolution of European modern art in the twentieth century, featuring major works by six esteemed modern masters, including Pablo Picasso, Paul Klee, Henri Matisse, Alberto Giacometti, Paul Cézanne, and Georges Braque.

Compared to UCCA Beijing, UCCA Edge enjoys the advantage of convenient transportation, with direct access to the exhibition hall from the subway station. This accessibility makes visiting the museum a seamless experience, even during white-collar workers' lunch breaks. Leveraging these benefits, UCCA Edge aims to attract a diverse audience and play a significant role in Shanghai's thriving art scene, all while maintaining a strong connection with its Beijing counterpart.

**参观指南**

◎ 上海市静安区西藏北路88号盈凯文创广场2层
⊙ 周二至周日 10:00—19:00 （18:30最后入馆）
*周一闭馆，开放时间或因特殊展览或活动有所变化

**Visitor Information**

◎ 2nd Floor, Yingkai Cultural Plaza,
88 Xizang North Rd., Jing'an District, Shanghai
⊙ Tuesday—Sunday 10:00—19:00 (last admission at 18:30)
Closed on Mondays. Opening hours may vary for special exhibitions or events

"托马斯·迪曼德：历史的结舌"展览现场，2022
Thomas Demand: The Stutter of History, 2022
图片提供：UCCA尤伦斯当代艺术中心 Courtesy UCCA Center for Contemporary Art

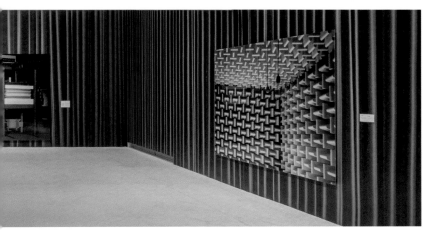

明当代美术馆外景　Exterior view of the Ming Contemporary Art Museum
图片提供：明当代美术馆　*Courtesy Ming Contemporary Art Museum*

# 13
# 明当代美术馆
# Ming Contemporary Art Museum

空间改造设计师:高士名,邱志杰 | 开馆时间:2015年
Space Renovation Designers: Gao Shiming, Qiu Zhijie | Opening Year: 2015

可以说,这是上海唯一一家专注于当代艺术中表演艺术门类的艺术机构。

Setting itself apart from other contemporary museums in Shanghai, McaM places a strong emphasis on promoting contemporary art activities rooted in visual performance.

上海明当代美术馆是由上海明园集团资助的非营利当代艺术中心。明园集团是国内较早介入当代艺术的房产企业,创办艺术机构,收藏艺术品,并积极参与具有公共性的艺术事件,比如承办2005年上海国际城市雕塑双年展。上海明当代美术馆创办后,除了沿袭明园既往对中国当代艺术的关注外,也做了更国际化的叙事拓展,比如2018年艺术家托马斯·赫赛豪恩在中国的首次个展。

该馆成立后,持续推出以视觉表演为基础的当代艺术活动,可以说,这是上海唯一一家专注于当代艺术中表演艺术门类的艺术机构。美术馆空间由原上海造纸机械厂的工作车间改造而成,特有的U字形空间的下沉式舞台架构,为各类表演提供了完美舞台。实验剧场、多媒体展演,与表演概念相链接的装置、绘画、雕塑、录像、声音、诗歌、舞蹈等活动,以及相关系列讲

The Ming Contemporary Art Museum (McaM) in Shanghai is a non-profit contemporary art center sponsored by the Ming Yuan Group, a real estate enterprise that has been actively involved in China's art scene since its early days. Over the years, the group has established art institutions, amassed art collections, and hosted art events, including the 2005 Shanghai International Urban Sculpture Biennale. McaM was inaugurated in 2015, continuing the group's commitment to Chinese contemporary art while embracing the international art world. Notably, it hosted Thomas Hirschhorn's first solo show in China, titled "Re-Sculpt," in 2018.

Housed in a transformed former workshop of the Shanghai Paper Machine Factory, McaM preserves the original industrial architecture with its inherent simplicity. It features a U-shaped space with a sunken stage, covering an expansive area of 3,500 square meters. The interior space is divided into two floors, boasting a ceiling height of 15 meters, catering to the production needs of various experimental theaters, contemporary art exhibitions, and performances.

Setting itself apart from other contemporary museums in Shanghai, McaM places a strong emphasis on promoting contemporary art activities rooted in visual performance. With a specific focus on experimental theater, multimedia exhibitions, and performances, as well as installations, paintings, sculptures, videos, sounds, poems, and dances, the museum aims to foster profound experimentation and interaction that transcend these diverse artistic domains.

McaM showcases a wide range of performances, featuring renowned pioneers of China's experimental theater scene like Mou Sen and Zhang Xian, as well as international artists such as the Japanese group Maywa Denki and Daniel Belton, a multi-talented artist, filmmaker, choreographer, and dancer from New Zealand. Belton specializes in visually capturing the movements of dancers' bodies and transforming them through sound and new media technology. The museum also facilitates collaborations between multimedia artists like Ulf Langheinrich, who actively participates in major art museums and electronic music festivals worldwide, and young Chinese dancers.

What distinguishes performances at McaM from those in traditional venues is the perspective of viewing performances not merely as commodities but as artistic pieces within the exhibition framework of a museum. In September 2018, McaM organized a four-day event called "Staging Alterity 2018," featuring around 20 artists and troupes from both domestic and international backgrounds. The event included 14 plays, along with speeches and a closed-door workshop. Through such projects, McaM aims to explore the possibilities of performances within the context of a museum, pushing the boundaries of artistic expression.

座、工作坊,让这座3 500平方米、拥有15米挑高空间的美术馆成为立体、多元、动态的剧场。这里的演出涉及表演艺术的各个面向:从国内的先锋戏剧代表牟森、张献,到带有工业、工厂元素的日本演出厂牌"明和电机";从擅长利用新媒体技术将舞者身体的律动与声音变幻的痕迹视觉化,搭建融合身体与科技"影像舞台"的新西兰新媒体艺术家丹尼尔·贝尔顿,到活跃于世界各大美术馆与电子音乐节的多媒体艺术家、作曲家乌尔夫·朗因海里希与中国青年舞者的合作,等等。

虽然都是表演,但相较于演出场所不同的是,美术馆不会将表演视为商品。表演一旦进入美术馆系统,会如同艺术作品一般,在展览的框架下呈现;只是,作为表演艺术的作品是动态的。2018年中秋假期,明当代美术馆举办为期四天的公共项目"他者的舞台",汇聚亚非拉及欧洲年轻表演力量在艺术领域的探索,探讨表演在其现场的无限拓展可能。

工人、劳动与工人社区也是明当代一条动人的叙事脉络,这与该馆原址作为工厂车间的身份相呼应。2021年"五周年文献回顾+邀请展:再回现场"以"生产行动"作为序幕,将身为艺术家的身体劳作与车间工人伴随机器的劳作交叠显像。"社区剧场"是明当代一项面向大众、旨在建立艺术家与居民社区沟通的持续项目,曾以20世纪工人阶级的文化休闲生活的场所——工人文化宫为灵感,展开对当下网络社会劳动者生活与文化状况的研究。

"明和电机:超常识机械"展,表演现场,2016
"Maywa Denki: Nonsense Machine" live performance, 2016
图片提供:明当代美术馆 *Courtesy Ming Contemporary Art Museum*

"后感性：恐惧与意识"展览现场，2017
Post-sense Sensibility: Trepidation and Will, 2017
图片提供：明当代美术馆 *Courtesy Ming Contemporary Art Museum*

上图:"托马斯·赫赛豪恩:重塑"展览现场,2018;下图:杨·罗威斯和尼德剧团的长时表演《父之屋》,2016
Above: Thomas Hirschhorn: Re-Sculpt, 2018; Below: Jan Lauwers' opera staging "L'incoronazione di Poppea", 2016
图片提供:明当代美术馆 *Courtesy Ming Contemporary Art Museum*

The museum's narrative intertwines with the themes of labor and worker communities, resonating with the site's original identity. In the opening chapter of the Fifth Anniversary Documentary Retrospective and Invitational Exhibition in 2021, titled "Production Action," the labor of artists, embodied in their creative work, was juxtaposed with the labor of factory workers alongside machinery.

McaM's community project, "Community Theater," draws inspiration from the historical significance of the "Worker's Cultural Palace," which symbolized the cultural and recreational activities of the working class in China during the 20th century. This project delves into research on the lives and cultural conditions of contemporary laborers in the context of a networked society. Its primary objective is to foster communication and interaction between artists and local residents, creating a platform for shared experiences and perspectives.

**参观指南**

- 上海市静安区永和东路436号
- 周二至周日 10:00—18:00（17:30停止入园）
  周一闭馆

**Visitor Information**

- 436 East Yonghe Rd., Jing'an District, Shanghai
- Tuesday—Sunday 10:00—18:00 (last admission at 17:30)
  Closed on Mondays

这座由原上海造纸机械厂的工作车间改造而成的美术馆，拥有3 500平方米和15米挑高空间，已经成为了一个立体、多元、动态的剧场
Transformed from a former workshop of the Shanghai Paper Machine Factory, McaM boasts an area of 3,500 square meters with a ceiling height of 15 meters, allowing it to evolve into a three-dimensional, diverse, and dynamic theater

图片提供：明当代美术馆 *Courtesy Ming Contemporary Art Museum*

明珠美术馆标志性的椭圆形外观 The elegant oval shape of the Pearl Art Museum
图片提供：明珠美术馆 *Courtesy Pearl Art Museum*

# 14
# 明珠美术馆
# Pearl Art Museum

建筑师：安藤忠雄建筑事务所 ｜ 开馆时间：2017 年
Architects: Tadao Ando Architect & Associates ｜ Opening Year: 2017

> 明珠美术馆是商场里的美术馆，但在空间上并无局促感，建筑空间设计由安藤忠雄操刀，让其在空间及美学上都获得了独立属性。
>
> The museum itself finds its home on the eighth floor within a shopping mall. It forms part of a two-story complex masterfully designed by the renowned Japanese architect Tadao Ando.

上海西郊是成熟的大型生活区，有建于 20 世纪 50 年代的上海动物园及 90 年代的大型居民住宅区，也有建于 2010 年后的超大摩登商业综合体。上海明珠美术馆是商场里的美术馆，但并无局促感：一是因为商业体的庞大体量，二是由于美术馆位于顶层，有足够的舒展空间，加上建筑空间设计由安藤忠雄操刀，让其在空间及美学上都获得了独立属性。位于 8 层的美术馆和 7 层的书店是一体的，是闪耀着银色金属光泽的蛋形建筑。连通 7 层与 8 层的"心厅"挑高 10 米，有着安藤忠雄建筑标志性清水混凝土浇筑的墙面。美术馆的 logo，一个类似水滴的椭圆形，根据安藤忠雄画的蛋形手稿变形得来。

虽偏居一隅，与上海的几处艺术带均相隔较远，但该馆还是走出了自己的特色。除了与欧洲博物馆、收藏机构合作的中国首展，如捷克国宝级艺术家慕夏的展览、"维克多·雨果：天才的内心"外，还有颇具特色的独立策展。

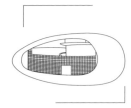

The Pearl Art Museum is nestled in the western suburbs of Shanghai, an area characterized by a blend of established developments. Here, the Shanghai Zoo, established in the 1950s, coexists with large-scale residential blocks from the 1990s and modern commercial complexes constructed after 2010. The museum itself finds its home on the eighth floor within a shopping mall. It forms part of a two-story complex masterfully designed by the renowned Japanese architect Tadao Ando. The complex also houses a bookstore called "Xinhua Culture & Creativity—Light Space" on the floor below. The architecture of the complex features an elegant oval shape, with a grand hall connecting the core areas of both floors. This hall boasts a soaring 10-meter ceiling and showcases Ando's signature poured concrete walls.

Despite its location within a shopping mall, the museum defies any sense of confinement or limitation. This is attributed to several factors, including the spaciousness of the complex, the museum's position on the top floor, and Ando's skillful manipulation of space. The overall design creates an atmosphere of openness and expansiveness, facilitating an environment ideal for the appreciation and immersion in art.

The Pearl Art Museum is a non-profit private art institution resulting from a collaboration between Shanghai Xinhua Distribution Group and Red Star Macalline Home Group. It stands as the first art museum in China to merge the concepts of an "art museum + bookstore." Books play a significant role in the museum's exhibitions, adding a unique dimension to the artistic experience.

For example, the 2020 exhibition "Landscape and Books" explored the artistic qualities of books, using "books as art" as a starting point. It delved into the artistry contained within books, inviting the public to discover the stories, characters, and landscapes encapsulated within their pages.

Another noteworthy exhibition, "Encounter of Imaginations: Dialogue between *The Divine Comedy* and *Classic of Mountains and Seas*," held from November 2021 to February 2022 to commemorate the 700th anniversary of Dante's death, featured four rare copies of *The Divine Comedy* from the Encyclopedia of Tregani Academy in Italy and two rare Ming Dynasty editions of *Classic of Mountains and Seas* from the Shanghai Library's collection. These precious works were displayed within a specially designed space. The exhibition also invited five Italian artists and seven Chinese artists to create their interpretations of these literary classics, fostering a meaningful dialogue between the two countries.

Additionally, the museum collaborates with international art institutions to bring a diverse range of art to the Shanghai audience. Exhibitions such as "Mucha" and "Victor Hugo: Dans L'intimité Du Génie" in 2019 exemplify these collabo-

作为由上海新华发行集团与红星美凯龙家居集团合作创建的美术馆，美术馆与书店共生成为该馆的最大特色，书籍如基因一般编织在展览中。比如，在但丁逝世700周年之际推出的"想象的相遇：《神曲》对话《山海经》"展中，美术馆特别规划空间，展出来自意大利特莱加尼百科全书学院的4本14世纪珍贵《神曲》手抄本还原，以及上海图书馆馆藏明代《山海经》珍稀刻本2部。"风景与书"展，更是从"书籍之为艺术"出发，让大众看到了书籍中的故事、人物与风景。

除了书籍，设计——准确讲是"生活化的设计"，也是明珠美术馆的一条策展脉络。这与该馆所在场域也契合。开馆展由建筑师安藤忠雄的设计展打响，2022年设计活动家长冈贤明在此系统演绎了他所主张的"长效设计"理念。

"没有足够的距离：跨界潮流艺术展"展览现场，2021
Manuque de Recul: Interdisciplinary Trends in Art, 2021
图片提供：明珠美术馆 *Courtesy Pearl Art Museum*

"读书行路:《路易威登游记》艺术展"
展览现场, 2018
Reading Walking: Louis Vuitton Travel Book, 2018

图片提供：明珠美术馆 *Courtesy Pearl Art Museum*

"风景与书：明珠美术馆两周年庆典展"
展览现场, 2020
Landscape and Books: Pearl Art Museum's 2nd Anniversary Celebration Exhibition, 2020

图片提供：明珠美术馆 *Courtesy Pearl Art Museum*

"想象的相遇：《神曲》对话《山海经》"
展览现场, 2021
Encounter of Imaginations: Dialogue between *The Divine Comedy* and *Classic of Mountains and Seas*, 2021

图片提供：明珠美术馆 *Courtesy Pearl Art Museum*

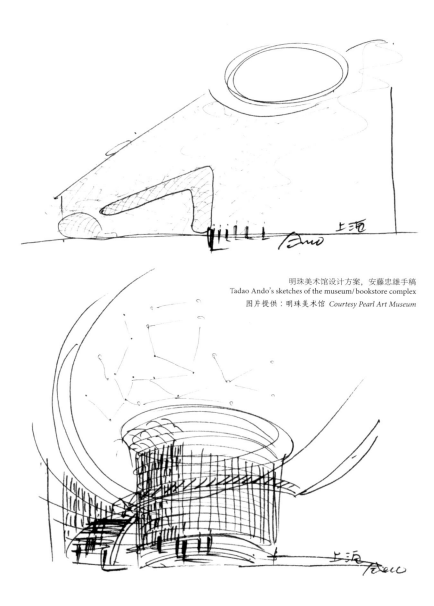

明珠美术馆设计方案,安藤忠雄手稿
Tadao Ando's sketches of the museum/bookstore complex
图片提供:明珠美术馆 *Courtesy Pearl Art Museum*

rations, offering visitors the opportunity to engage with renowned international artworks and immerse themselves in different artistic styles and expressions.

Design serves as another prominent curatorial theme at the Pearl Art Museum. The museum has presented exhibitions that explore various aspects of design and highlight the works of acclaimed designers. In its inaugural exhibition, "Tadao Ando: Leading" (December 2017–May 2018), the museum showcased the design philosophy and accomplishments of the esteemed architect.

In 2022, the museum featured an exhibition titled "Long Life Design: Thinking and Practice," presented by Japanese design activist Kenmei Nagaoka. Through this exhibition, Nagaoka conveyed his concept of "long life design," emphasizing sustainability, longevity, and responsible design practices. The exhibition offered an opportunity to contemplate the role of design in creating a more sustainable future.

## 参观指南

⦿ 上海市闵行区吴中路1588号爱琴海购物公园8楼
🕐 周二至周四 10:00—19:00（18:00停止入场）
　周五至周日 10:00—22:00（21:00停止入场）
　每周一闭馆

## Visitor Information

⦿ 8F, Aegean Place, 1588 Wuzhong Rd.,
　Minhang District, Shanghai
🕐 Tuesday–Thursday 10:00–19:00 (last admission at 18:00)
　Friday–Sunday 10:00–22:00 (last admission at 21:00)
　Closed on Mondays

心厅位于"光的空间"核心位置，其椭圆形结构模型现陈列于8楼明珠美术馆入口处
The 10-meter high hall is at the heart of the two-story "Light Space,"
its elegant oval-shaped structural model is exhibited at the entrance of the museum

图片提供：明珠美术馆 *Courtesy Pearl Art Museum*
摄影：李雪峰 *Photo by Li Xuefeng*

没顶美术馆启用了20世纪70年代崇明农场的原生态厂房
The MadeIn Art Museum has been established in the original ecological factory buildings of the Chongming Farm from the 1970
图片提供：没顶美术馆 *Courtesy MadeIn Art Museum*

# 15
# 没顶美术馆
# MadeIn Art Museum

开馆时间：2022 年　Opening Year: 2022

在地性一直是艺术创作高呼的原则，在这里，没有口号，只有润物细无声的实践。

The principle of locality is meticulously upheld within the museum's premises.

从上海市区开车两小时，穿过隧道和长江大桥，登陆崇明岛，在崇明岛的最西头，是没顶美术馆的所在。与馆长、艺术家徐震一贯的大胆"逾矩"作风一致，没顶美术馆也超越了人们对美术馆的既有印象。美术馆空间的前身为养鸡场的鸡舍，始建于 20 世纪 70 年代。美术馆周边是原生态的乡野、农田，而非产业化的休闲度假农庄，没有城市消费主义的一切痕迹。展厅保留了原有空间结构，长长的过道仿若 T 台，引人步入，观看一件件作品；而这个"T 台"又天然地将展场划分成左、右两个区域。原场房是 20 世纪集体主义农场的布局，一排排整齐地分布在道路两侧，这 8 座平房现在是 8 个展厅：有的展厅屋顶没有瓦片，硕大的雕塑作品穿越梁栋"长"向天际，周身还缠绕着野生植物的绿色枝蔓；有的展厅内长着茂盛树木，行为艺术家进行表演时会拨开树枝，仿若身处森林。

The MadeIn Art Museum is nestled in Lvhua Town, situated in the western region of Chongming Island. This alluvial island, located at the mouth of the Yangtze River, offers a unique setting for contemporary art enthusiasts. A tranquil two-hour drive from downtown Shanghai, visitors are greeted by a picturesque landscape of tree-lined paths leading to the island's westernmost point. Here, a series of eight bungalows, arranged neatly on either side of the road, stand as a testament to the area's former chicken farm from the 1970s. Despite years of abandonment, the surroundings remain untouched by urban consumerism, offering pristine rural vistas and farmlands.

The original farm buildings, reflecting the collective farming style of the 20th century, have been thoughtfully repurposed into eight exhibition halls, with their original structures intact. Some halls still feature untiled roofs, while large sculptures soar high above, entwining with lush green branches and wild plant vines. In other halls, verdant trees grow inside, and performance artists may find themselves parting branches as they immerse themselves in their acts, as if within a jungle.

Chongming Island, predominantly reclaimed land from the 1970s, spans over 40,000 acres and serves as a remarkable example of socialism. Xu Zhen, the artist and curator of the museum, refers to his experiences on the island as an "artistic collective life." To him, the island acts as a production base, an art creation camp, and a collective organization liberated from individual influences. Xu's artistic involvement with the island extends beyond the establishment of this museum. On this occasion, he brought over a dozen sculpture works, including oversized red high heels emerging within a house ["Eternal Life (Greek Column, High Heels)"] and a glass case by a small river housing a "dinosaur" specimen.

The principle of locality is meticulously upheld within the museum's premises. At the entrance, visitors are greeted by a towering 20-meter-high installation titled "How to Stand Up a Huge Column Sculpture" by artist Feng Zhixuan. This artwork incorporates materials such as seashells, volcanic rocks, and coral bones, which the artist poured into a nearby ditch and erected once they solidified. Xu Zhen adds a touch of humor by labeling the wrapped trash bins at the entrance with the exhibition name and participating artists' names.

In this serene environment, art is liberated from urban contexts and acquires a purer sense of appreciation. From the inaugural exhibition "We Borrow Dreams from Others, like Debts" (October 2022) to "Living a Performance Artist's Life: 2023 Performance Art Documentary Exhibition" (May–September 2023), the focus remains on artistic collectivism. As Xu Zhen expressed, "we simply want to engage in this game of artistic collectivism, and a museum like this adds pleasure to the game, training players' abilities and constantly increasing the challenge."

20世纪70年代,作为滩涂的崇明岛被填出了4万多亩沃土,成为社会主义垦荒史的典范。擅长波普主义的徐震,也在此"空投"了十几个雕塑作品,如在屋里长出的硕大红色高跟鞋,在小河边装有"恐龙"标本的玻璃箱等。

在地性一直是艺术创作高呼的原则,在这里,没有口号,只有润物细无声的实践。美术馆入口处一座高达20米的装置作品,翻模自旁边的小水沟,艺术家冯至炫把所有材料都倒到这条小河里面,凝固之后竖立起来;而作品里的材料均来自海——牡蛎壳、火山石、珊瑚骨头之类。展陈上也能看到徐震一贯的幽默,美术馆入口的垃圾桶被包裹着,上面印着展览名称及参展艺术家的名字。

在乡野,一面认真地做艺术展览,从开馆展"我们从别人那里借梦想,像债一样"到行为艺术文献展"活成行为艺术家";一面认真地饲养动物,小马、小羊之类。艺术从都市情境中被释放出来,获得了更为纯粹的趣味。"我们其实就是想要去玩这种艺术集体主义生活的游戏,而一个这样的美术馆,可以给这种游戏增加乐趣,训练玩家的能力并不断提高挑战难度。"徐震说。

徐震®作品《永生》,长期陈设
"Eternity" (Greek Columns, Heels) by Xu Zhen®, permanent display
图片提供:没顶美术馆 *Courtesy MadeIn Art Museum*

"我们从别人那里借梦想，像债一样"
展览现场，2022
We Borrow Dreams from Others,
Like Debt, 2022

图片提供：没顶美术馆
*Courtesy MadeIn Art Museum*

徐震®作品《永生：新40403石佛像，阿芙罗狄蒂》，长期陈设
"Eternity: New 40403 Stone Statue, Aphrodite Holding Her Drapery"
by Xu Zhen®, permanent display

图片提供：没顶美术馆
*Courtesy MadeIn Art Museum*

徐震®作品《永生：河北省博物馆唐朝白石菩萨立像、青州龙兴寺北齐卢舍那法界人中像、曲阳修德寺石菩萨像、阿法埃娅神庙》，长期陈设
"Eternity: Tang Dynasty Bodhisattva of the Hebei Province Museum, Northern Qi Losana Buddha of the Longxing Temple, Bodhisattva of the Xiude Temple, West Pediment of the Temple of Aphaia"
by Xu Zhen®, permanent display

图片提供：没顶美术馆
*Courtesy MadeIn Art Museum*

任莉莉作品《梦》,"我们从别人那里借梦想,像债一样"展览现场,2022
"The Skeleton of a Mattress" by Ren Lili, We Borrow Dreams from Others, Like Debt, 2022

图片提供:没顶美术馆 *Courtesy MadeIn Art Museum*

俯瞰图　Aerial view
图片提供：没顶美术馆 *Courtesy MadeIn Art Museum*

**参观指南**

◎ 上海市崇明区绿华镇环湖西路
◔ 周三至周日 *10:00—16:30*（需预约）
  节假日不开放

### Visitor Information

◎ *Huanhu West Rd., Lvhua Town, Chongming District, Shanghai*
◔ *Wednesday—Sunday 10:00—16:30 (reservation required)*
  *Closed on national holidays*

徐震®作品，《新：拉奥孔》，长期陈设
"New: Laocoön" by Xu Zhen®, permanent display
图片提供：没顶美术馆 *Courtesy MadeIn Art Museum*

# Afterword

In recent years, "citywalk" has been steadily gaining popularity. An article titled "Chinese People Who Don't Excel at Going Crazy Are Secretly Embracing Citywalk" was recently published on the "Phoenix Weekly" WeChat account on July 5, 2023. Subsequently, it was reposted by the InsDaily WeChat account, and the article has garnered a readership of nearly 150,000.

Citywalk is a term of foreign origin, but in more authentic Chinese terms, it is simply known as "遛弯" (*liuwan*), which refers to strolling aimlessly or wandering around a city. Citywalk encompasses activities such as exploring shops, taking photos for social media check-ins, or simply going for a walk without a specific destination in mind. It has evolved into a symbol of a healthy lifestyle embraced by modern urban dwellers, who use the relaxed and casual nature of citywalk as a way to counter the pressures of societal competition.

The "CityWalk Book Series" was first introduced by Tongji University Press in 2012, pioneering the concept of citywalk in the Chinese publishing industry. Over the past decade, the series has published 17 titles, and these pocket-sized "rainbow books" have become a renowned publishing brand of popular books under Tongji University Press. This accomplishment is a testament to the hard work and dedication of generations of editors, and it is now our responsibility to carry on this legacy.

Simultaneously, over the past decade, cities as organic entities have been constantly evolving. The rise of self-media and AI has brought about transformative changes in content creation, knowledge organization, and dissemination. Generation Z, comprising those born after 2000, and even after 2010, now constitutes the primary target audience for popular books, and books are no longer their main source of information and learning. Faced with this dynamic landscape, we are prompted to question the significance of publishing in the digital age. This is a question that resonates with everyone involved in the industry. It is with these contemplations in mind that we embarked on the planning and launched the "Shanghai Art Museums' CityWalk Program."

# 编后记

近年来,"城市行走"逐渐火了起来。"凤凰 Weekly"微信公号于 2023 年 7 月 5 日发布了一篇名为《不擅长发疯的中国人,都在偷偷 citywalk》的文章,之后 InsDaily 微信公号进行了转载,其阅读量已经近 15 万。

Citywalk 是个"洋名",用更地道的中国话说其实就是"遛弯",探店能叫 citywalk,打卡能叫 citywalk,出门散个步也能叫 citywalk,无目的地瞎逛更是 citywalk……它已经成为现代都市人的一种健康生活方式的象征,用 citywalk 的随性来抵御内卷的压力。

同济大学出版社的"城市行走书系"开始于 2012 年,可以说是在出版界首次引入了 citywalk 这个概念。十年间,陆陆续续出版了 17 个品种,这套口袋本的"彩虹书"也成为同济大学出版社大众类图书的重要品牌,得到业内一致好评。这是一代又一代编辑辛勤付出、默默努力的成果,我们要做的是传承。与此同时,十年间,城市作为一个有机体不停地在自我迭代,自媒体和 AIGC 的发展使得内容创作、知识组织和传播发生了质的改变,Z 世代(零零后甚至壹零后)作为大众图书的主要读者对象,图书已经不再是他们主要的信息获取来源和学习途径了,面对这种变化,出版的价值在哪里?我们相信,这是所有人都会问的问题,正是带着这些思考,我们开始了"行走上海美术馆"的策划和创作。

在策划之初,一切都没有那么清晰,只是运用了现有的资源,跟着感觉走而已。机缘巧合地结合了同济大学设计创意学院环境设计专业的大二课程,与一批零零后一起调研了上海的 5 个美术馆,和这些学生的接触,开启了我们全新的视角,也让我们重新审视了 citywalk 的意义。随后,

At the outset of the planning process, everything was far from clear, and we relied on our instincts while making the most of available resources. By a stroke of serendipity, we collaborated with second-year students from the Environmental Design Department of the College of Design and Innovation at Tongji University. Together with a group of post-2000 students, we conducted research on five art museums in Shanghai. This collaboration brought about a fresh perspective, leading us to reevaluate the significance of citywalk. Subsequently, we embarked on visits to the 15 art museums featured in this book and conducted interviews with curators, architects, and other key figures, which gradually sharpened our understanding and insights.

*Roaming Shanghai's Art Museums: A CityWalk Exploration* goes beyond the conventional book format, offering an open platform for discussion and interaction. It takes on diverse manifestations, serving as a book, a course, a project, a series of events, a collection of cultural and creative products, a series of vlogs, a new media platform, and more. Regardless of its external appearance, its essence remains rooted in uniting like-minded individuals for captivating explorations, citywalk adventures, design appreciation, art discussions, and the sharing of life's experiences. This embodies the authentic spirit of citywalk.

We express our heartfelt gratitude for the generous support and active participation from the art museums, teachers, and students involved in "Shanghai Art Museums' CityWalk Program." Special thanks go to CPC Secretary Chen Yan, Professor Sun Jie and Associate Professor Ding Junfeng of the College of Design and Innovation at Tongji University for providing us with a cooperative platform that has allowed us to envision a future driven by design.

We would also like to extend our appreciation to the following students who have been instrumental contributors to this book and its related content: Sun Yihan, Fu Yutong, Wu Yue, Ni Mingjie, Zhang Hanchi, and Su Shimiao. Their dedication and efforts have played a significant role in shaping the success of this endeavor. Thank you all for being an essential part of this remarkable journey.

Without further ado, you are warmly invited to join us in the enchanting world of Shanghai's art museums. Let's meet and immerse ourselves in the captivating beauty of art together! See you at the museums!

*Yuan Jialin*
*Initiator of "Shanghai Art Museums' CityWalk Program"*

作者潘丽陆续走访了本书所收录的 15 个美术馆，采访了一些馆长、建筑师和策展人，我们的思路也慢慢变得清晰起来。

在读者手上的《行走上海美术馆》是一本书，但其实我们更想让大家将其理解为一个开放的话题平台，它可以是一本书、一门课、一个项目，也可以是一系列活动、一堆文创、一串 vlog、一个新媒体平台……不管它的外在形式是什么，其本质无非就是聚集一些志同道合的人，做一些有趣的事儿，逛逛城市、看看设计、聊聊艺术、分享下生活，而这正是 citywalk 的精神所在。

非常感谢支持和参与"行走上海美术馆"的美术馆、老师们和同学们，特别要感谢同济大学设计创意学院的陈燕书记和丁峻峰副教授所提供的合作平台，还要感谢孙捷教授的启发和指导，让我们看到了由设计驱动的共创未来，也要感谢以下主要参与《行走上海美术馆》及相关内容创作的同学，他们是孙弋涵、付雨彤、吴越、倪铭杰、张寒驰、苏诗淼。

好了，不多说了，让我们在上海的美术馆相遇吧！

袁佳麟
"行走上海美术馆"发起人

图书在版编目（CIP）数据

行走上海美术馆：汉、英／潘丽著. -- 上海：同济大学出版社, 2023.8
ISBN 978-7-5765-0908-3

Ⅰ.①行… Ⅱ.①潘… Ⅲ.①美术馆 - 建筑艺术 - 上海 - 汉、英 Ⅳ.①TU242.5

中国国家版本馆CIP数据核字(2023)第149941号

# 行走上海美术馆
# ROAMING SHANGHAI'S ART MUSEUMS
## A CityWalk Exploration

潘丽 著　Pan Li

| | |
|---|---|
| 出 品 人 | 金英伟 |
| 责 任 编 辑 | 袁佳麟 |
| 特 约 编 辑 | |
| 英 文 翻 译 | 杨碧琼 |
| 编 辑 助 理 | 孙弋涵 |
| 责 任 校 对 | 徐逢乔 |
| 封 面 设 计 | 张微 |
| 装 帧 设 计 | 付雨彤 |
| 插 画 设 计 | 郝晟（封面） |
| | 吴越（内页） |
| 出 版 发 行 | 同济大学出版社　https://press.tongji.edu.cn |
| | 地址：上海市四平路1239号　邮编：200092 |
| | 电话：021-65985622 |
| 经 销 | 全国各地新华书店 |
| 印 刷 | 上海安枫印务有限公司 |
| 开 本 | 787mm×1092mm　1/32 |
| 印 张 | 6.625 |
| 字 数 | 148 000 |
| 版 次 | 2023年8月第1版 |
| 印 次 | 2023年8月第1次印刷 |
| 书 号 | ISBN 978-7-5765-0908-3 |
| 定 价 | 78.00元 |

本书若有印装质量问题，请向本社发行部调换
版权所有 侵权必究